General Preface to the Series

Recent advances in biology have made it increasingly difficult for both students and teachers to keep abreast of all the new developments in so wide-ranging a subject. The New Studies in Biology, originating from an initiative of the Institute of Biology, are published to facilitate resolution of this problem. Each text provides a synthesis of a field and gives the reader an authoritative overview of the subject without unnecessary detail.

The Studies series originated 20 years ago but its vigour has been maintained by the regular production of new editions and the introduction of additional titles as new themes become clearly identified. It is appropriate for the New Studies in their refined format to appear at a time when the public at large has become conscious of the beneficial applications of knowledge from the whole spectrum from molecular to environmental biology. The new series is set to provide as great a boon to the new generation of students as the original series did to their fathers.

1986

Institute of Biology
20 Queensberry Place
London SW7 2DZ

Preface to First Edition

There is a rumour prevalent in the common rooms of the medical faculties that headmasters tend to keep medicine for their less brilliant scholars, and that the best scientific brains are advised to go into mathematics, chemistry or physics. It is the hope of the author that this book may go a little way to remedying this view, for it should be made clear to young people that medicine is a most exciting subject intellectually and that with the biological explosion upon us there are many problems in medicine on the verge of being solved and the discoveries are likely to be made by those who see their subject in relation to general biology.

Various authors and editors have kindly given me permission to reproduce figures and tables from their publications, and these are all acknowledged where they appear. If the work is included in the list of references at the end, or in the bibliography, the full reference is given under the figure or table. I should like particularly to express my indebtedness to Mr Per Saugman and to Messrs Blackwell for their kindness in allowing me to include parts of my book *Genetics for the Clinician*.

Liverpool, 1970   C. A. C.

Preface to Second Edition

It gives me great pleasure to know that a second edition of this book is called for. *Post hoc* is not necessarily *propter hoc*, but I should be less than human if I did not rejoice in the fact that the young are now clamouring for places in the medical schools and medicine is seen to be a discipline satisfying to the most gifted scientific scholars.

I have tried to bring the second edition up to date but where basic principles are involved it has often seemed better to use the older, well-tested examples. I am most grateful to Dr D. A. Price Evans for advice on this principle in the Pharmacogenetics chapter.

Liverpool, 1977   C. A. C.

Preface to Third Edition

The scene has changed again; now there are too many doctors and technology (so it is alleged) is triumphing to the detriment of patient care. Do not believe a word of this. True, there are many advances in the management of disease, but behind the scenes, how to treat a particular patient creates much agonizing since the options are now greater than ever.

One aspect of old age is that it becomes increasingly difficult to understand jargon and so I have taken particular care about Genetic Engineering. I hope in Chapter 12 of this new edition that I have written an account which the really uninitiated can grasp. It has taken me weeks of toil and trouble and even now many of the ingredients of the witches' brew still defeat me.

The rest of the book deals with the 'old' genetics and it is important for the new generation to realize that what is old is not necessarily either out of date or untrue. The emphasis in selecting the topics is on interest rather than comprehensiveness.

I am most grateful to Dr D. A. Price Evans, Director of Medicine at the Riyadh Armed Forces Hospital, Saudi Arabia, for the very great help he has given me in all the chapters. He should really have been a co-author but preferred to leave my style alone, and so all I can do is to thank him very much indeed and to stress that all the errors are my own responsibility.

I am most grateful to Mr K. R. C. Neal, Head of Biology, Manchester Grammar School, for useful discussions, to my wife for being at my beck and call night and day and to Edward Arnold for much cooperation and forbearance.

Liverpool, 1987   C. A. C.

Contents

Contents

1

Dominant and Recessive Inheritance

1.1 Dominant inheritance: Huntington's chorea as an example of a rare disorder

Huntington's chorea (HC) is an inherited disease characterized by involuntary muscular movement and progressive mental deterioration. The age of onset is usually about 35 years so that the majority of those affected can produce a family before they are aware of their plight. The disease is transmitted by an autosomal dominant gene (Fig. 1.1), so both sexes are equally affected and, moreover, because penetrance (see glossary) is complete the disorder never skips a generation. It is rare − one estimate in this country is five cases per 100 000 of the population − and so affected individuals are highly likely to be heterozygotes (see below for explanation of this point). The mutant gene at the HC locus is probably on chromosome 4 and it is now possible in certain families to tell which individuals are at risk (see Chapter 12).

The disease, introduced into North America by two Suffolk immigrants in 1630, derives its name from the US doctor who first described it in 1872. Fraser Roberts (1973) writes 'the boy George Huntington, driving through a wooded lane in Long Island while accompanying his father on professional rounds, suddenly came upon two women, mother and daughter, both tall, thin, almost cadaverous; both bowing, twisting, grimacing, so that he stared in wonderment, almost in fear. The memory was as vivid more than fifty years later, long

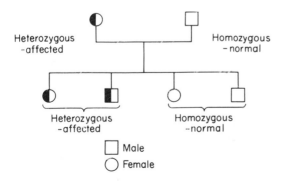

Fig. 1.1 Pedigree of Huntington's chorea (HC).

er he had translated into fact the youthful resolve, born that day, to make
norea the subject of his first contribution to medical science: a resolve which
ed him into many a home where the bearers of the gene waited with stern
Calvinistic stoicism for the dreadful fate that Providence had meted out to
them.'

Affected individuals will almost invariably be heterozygous for the condition
because their parents will have married normal partners. Though theoretically
an affected × affected mating *could* take place and produce a homozygote, this
would probably be lethal, and the gene is so rare that it is most unlikely to
occur. By contrast, where a trait is common, such as the various ABO blood
types, it is readily possible to have the homozygote, e.g. blood group OO or AA.

1.2 The problem of controls in assessing the mutation rate in HC

A matter of considerable interest is the reason for the persistence of the disease.
'Recurrent mutation' is the obvious answer, but another explanation is that,
since heightened sexual desire (increased libido) is one of the early symptoms,
affected individuals will have more children than their unaffected siblings
(brothers and sisters) and this in fact has been found to be so. The mutation
rate, therefore, could be extremely low (or even non-existent) since, on the
assumption that unaffected sibs behave as normal individuals, the biological
fitness (see glossary) is higher than unity in the patients. However, to use the
unaffected sibs may not be the right comparison because, since they are aware
of the inherited nature of their disease, they may marry late or limit their
families. In fact, where the fitness of HC individuals is compared with *normal*
people it is well below unity (0.81) and if these are the right controls (as seems
likely) then a higher mutation rate must be invoked. It might be thought that
this is only of theoretical importance but in an age of radiation hazards *any*
information about mutation rates is of great importance, and HC shows how
difficult it is to assess it − a value for biological fitness being necessary for the
calculation.

The situation in HC should be compared with that of duodenal ulcer (see
p. 46) and it will be clear that it is perfectly legitimate to draw opposite conclu-
sions about the best type of control in the two diseases.

1.3 Dominant inheritance: Dupuytren's contracture as an example of a common disorder

Few doctors have ever seen a case of HC whereas Dupuytren's disease is very
common and therefore of more general interest. Figure 1.2 shows the contracture in the hands (the feet may also be affected) which is the result of the formation of an abnormal type of connective tissue (collagen). The condition is
inherited as an autosomal dominant: non-familial cases are also reported but
this is probably because insufficient care has been taken in examining the rela-

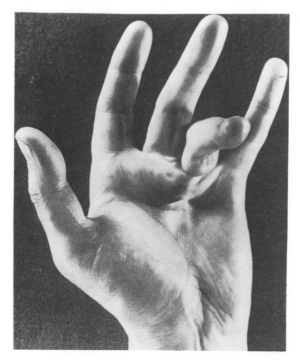

Fig. 1.2   Dupuytren's contracture. (From Hueston and Tubiana (1974), *Dupuytren's Disease*, Churchill Livingstone.)

tives. The condition may be very mild in young adults, is usually much less marked in women but all cases worsen with age and it has been stated that 25% of male old age pensioners have the condition. It is not surprising, therefore, that a host of associations has been found, e.g. with diabetes mellitus, alcoholism and epilepsy. An interesting piece of research would be to find out what the homozygotes look like and with such a common condition this could easily be done. The gene in double dose may simply result in a worse form of the disorder or possibly produce some other abnormality.

1.4   Recessive inheritance: (a) fibrocystic disease (FCD)

This disease, mainly affecting Europeans, is a generalized disorder of mucus secreting glands, particularly those in the pancreas, the intestines and the lungs. The mucus is more viscid than normal and, as a result, dried up secretions block the glands and their ducts so that they atrophy and become replaced by scar tissue. However, the cells in the pancreas which secrete insulin are not affected so diabetes does not occur. Another feature is that the sweat contains a higher concentration of sodium chloride than normal i.e. above 60 mmol/litre in infants or 100 to 110 mmol/litre in adults.

The disease is fairly common, occurring once in about 2000 births, and it accounts for between 1% and 2% of admissions to children's hospitals. The outlook for a sufferer is not good even with antibiotics and pancreatic extracts, many of the children dying of pneumonia, though a few survive to adult life and may have offspring.

The disease is an autosomal recessive condition, so that both sexes are equally affected and as a rule neither parent manifests the condition. Figure 1.3 shows that one in four individuals on average will be affected in a sibship where the disease occurs.

Possible explanations of the high frequency of fibrocystic disease

The fact that the disease is often met with makes the genetics of considerable interest and any of the following would explain the frequency:

(a) A high mutation rate – but this would have to be so high as, *a priori*, to be unlikely.

(b) Several different genes, each with its own mutation rate, might cause the condition which could vary in severity according to which of several alleles was responsible. In other words, the disease might be heterogeneous (as is often the case with other conditions) and possibly in support of this, the association with HLA types varies in different families (see p. 53).

(c) The heterozygotes, i.e. those individuals which carry one dose of the gene and which form 5% of the entire population (see p. 31 for the way in which this figure is calculated), might have an advantage, that is be biologically fitter, than the normal homozygotes. Thus the disease might constitute a polymorphic system (see p. 22) though what the advantage of the heterozygotes may be or may have been is unknown. Danks *et al.* (1965) investigated the family size of parents of children with fibrocystic disease, and they *did* find when investigating 144 of these grandparental couples, matching three different control groups with each grandparental pair, that there was a tendency for the grandparents of children with fibrocystic disease to have larger families.

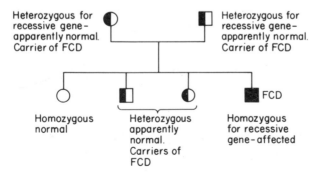

Heterozygous for recessive gene– apparently normal. Carrier of FCD

Heterozygous for recessive gene– apparently normal. Carrier of FCD

Homozygous normal

Heterozygous apparently normal. Carriers of FCD

Homozygous for recessive gene–affected

FCD

Fig. 1.3   Mating between two individuals heterozygous for the gene controlling FCD. Pedigree of fibrocystic disease (FCD).

Why grandparents? Parents of known fibrocystic disease patients might well limit their families, but in the generations before, heterozygotes would have been likely to marry normal people, and it is these who produced the bigger families than the controls.

The authors of the paper, however, put forward their results with great reservation, since the magnitude of the heterozygous advantage apparently shown is very much greater than would be needed to maintain the gene at a steady frequency in the community.

1.5 Detection of heterozygotes in fibrocystic disease

This still remains a problem, and the two following approaches have been made.

The sodium content of the sweat

It was thought at one time that the sodium content of the sweat in the heterozygotes might be helpful in identification. It is true that on average these have a somewhat higher value than do normal people yet the range is very great and there is much overlap. Furthermore, allowance is often not made for age. The sodium content of the sweat rises as one gets older and therefore comparison must be made between patients of similar age groups. Figure 1.4 demonstrates this point, where it will be seen that there is practically no difference in the sodium content of the sweat between parents of affected individuals, i.e. known heterozygotes, and that of normal controls *of the same age group.*

Linkage analysis

Since the basic gene effect remains unidentified the only available method for searching for its position is linkage analysis by exclusion mapping, using polymorphic DNA probes that have been mapped to a known chromosome region (see Chapter 12). In this way, using only samples from affected individuals and obligate heterozygotes a number of chromosome regions have been found *not* to be linked to fibrocystic disease. More recently (October, 1985), there have been advances indicating that chromosome 7 is involved and a new marker has been found, but not very close to the cystic fibrosis gene (see Chapter 12). The whole point of this genetic engineering exercise is to identify carriers of the gene or affected fetuses *in utero.*

1.6 Recessive inheritance: (b) phenylketonuria (PKO) (incidence 1 in 15 000 births)

The features are mental retardation, restlessness and anxiety, the muscles are hypertonic ('stiff') and there is often eczema and epilepsy. Affected children

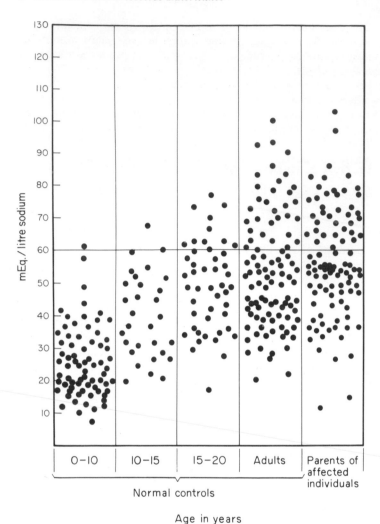

Fig. 1.4 Sodium content of sweat in relation to age. (Redrawn with permission from Anderson and Freeman (1960) 'Sweat Test' results in normal persons of different ages compared with families with fibrocystic disease. *Archives of Disease in Childhood*, **35**, 581−7.)

frequently have blue eyes and fair hair. The biochemical lesion is lack of a specific enzyme, phenylalanine hydroxylase, which normally converts phenyl-alanine into tyrosine. Everyone lacks the enzyme for about 2 days after birth but in phenylketonurics the synthesis never begins. Thus, 'the earlier the better' may not always be best for screening tests. Lack of tyrosine leads to diminution in melanin production, hence the characteristic fair hair of the patients. The diagnosis is made by a bacteriological inhibition test (the Guthrie test) performed on blood from a heel prick. The principle of the test is that the growth of *Bacillus subtilis* is inhibited by a chemical which in its turn is overcome by phenylalanine, thus allowing the growth of the bacillus i.e. if the bacillus grows the child has the disease.

A low phenylalanine diet often prevents symptoms developing but there are difficulties in deciding for how long the diet should be continued.

Mild cases may occur and phenylketonuric women bear children. A counsellor might be tempted to say that the risk of these children being affected is very small since it would only happen if the affected woman married a carrier. However, a woman with mild PKU may produce mentally retarded children because her phenylalanine has crossed the placenta, and although the affected children are not true PKUs this does not help them. Treated PKU women will increasingly pose the problem of whether dieting should be resumed during pregnancy.

There are still other problems. *The* feature of PKU is brain damage, but the primary enzyme defect lies in the liver and kidneys. Recently there have come to light variant forms of PKU; the result of defects in the metabolism of a co-factor of the missing enzyme, and a study of patients with these variants may perhaps teach us something about substances which are essential to normal brain development.

The antenatal diagnosis of PKU is in the process of being improved by the use of restriction fragment length polymorphisms and this principle is explained in Chapter 12.

2

Sex (X)-Linked Inheritance

2.1 Haemophilia as an example of sex-linked recessive inheritance

Haemophilia is a world-wide anomaly of blood coagulation (clotting) inherited as a sex-linked recessive, and is the disease which created such havoc in the royal families of Europe when introduced into them by the daughters and descendants of Queen Victoria (see Fig. 2.1). Haemorrhage, usually following injury but sometimes spontaneous, is the essential symptom. The bleeding is of the nature of a persistent, slow oozing which is out of all proportion to the extent of the injury; this can last for weeks and may lead to profound anaemia. Orthopaedic problems often occur due to recurrent haemarthroses (bleeding into the joints) but this can be greatly improved by aspiration.

The disease usually appears in early childhood or even in infancy, and it is due to the deficiency of one of the clotting factors of the blood (factor VIII, anti-haemophiliac globulin, or AHG) and is known as 'haemophilia A'. Its severity is very variable, yet within a given family it remains constant and this makes it likely that there are a number of alleles of the haemophilia gene. Also, there is a whole group of conditions which are the result of a deficiency of other blood clotting factors. The best known of these, Christmas disease (or 'haemophilia B') which was called after the patient in whom it was first described, is also a sex-linked recessive, and is due to deficiency of the Christmas factor (factor IX), which is distinct from AHG. A mixture of two equal parts of blood, one from a patient with haemophilia A and the other from a patient with Christmas disease clots normally, whereas separate samples show delayed clotting.

Another disorder of bleeding where the level of AHG is diminished is Von Willebrand's disease but this, unlike haemophilia, is inherited as an autosomal dominant. The disease is of considerable interest. Although factor VIII is again involved it is in a different way, there being defective synthesis of that part of the molecule which is concerned with the immunological factors needed for normal platelet function. The bleeding, therefore, is due to a different cause. The treatment is the same.

Returning to haemophilia it must be remembered that haemophiliacs may develop entirely unrelated diseases, and the writer has seen a fatal case of intestinal haemorrhage which was thought to be due to the haemophilia but was in fact secondary to a duodenal ulcer. In this connection, it is important to realize

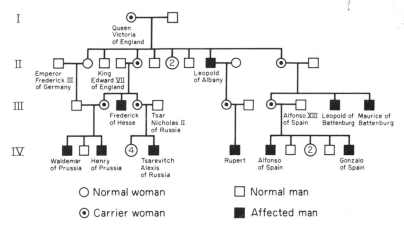

O Normal woman   □ Normal man

⊙ Carrier woman   ■ Affected man

Fig. 2.1 Sex-linked recessive inheritance. Pedigree of haemophilia in the royal families of Europe. All affected individuals trace ancestry to Queen Victoria of England, who undoubtedly was a carrier. Her father was normal, nothing suggests that her mother was a carrier, and therefore Queen Victoria seems to have received a *new* mutant allele from one of her parents.

It will be realized that many individuals in later generations are not included — among them our own royal family, who are quite free, being descended from an unaffected male. All Queen Victoria's children are entered in the pedigree. (By courtesy of Professor Curt Stern (1960) *Principles of Human Genetics*, Freeman.)

that haemophiliacs withstand operations quite well if they are suitably prepared with blood or (fresh) AHG beforehand.

The gene controlling haemophilia is situated on the X chromosome, and this is almost always what is meant by 'sex linkage' since Y linkage has very rarely been reported in Man, and several claims to its occurrence have not been substantiated on further investigation. It will be realized that since a woman is XX and a man is XY, an X-linked gene can be passed to either sex, whereas one on the Y could only go from father to son. (When a gene can only *express* itself in one sex it is said to be sex-limited or sex-controlled. This situation is described on page 12, and is quite different from sex-linkage.) Most, though not all, sex-linked genes in Man are recessive. Just as with the autosomes, crossing-over (see Chapter 8) can take place between one X and the other, but crossing-over between X and Y has only very rarely been reported, since there is a part of the X chromosome which does not pair with the Y.

It will readily be understood that since haemophilia is controlled by a sex-linked recessive gene, female carriers of the gene will never be affected unless they are homozygotes. This is because females have two X chromosomes and the normal gene on their second X will prevent the haemophilia gene from expressing itself — they will however pass on equal numbers of normal and abnormal X chromosomes to their offspring of both sexes, so that the offspring of carrier females will consist of half normal and half affected sons and of half normal and half carrier daughters. Owing to the rarity of the gene, females are most unlikely to be homozygotes and hence haemophiliacs, but cases have been reported and some patients have survived childbirth and borne haemophilic sons. Males, on the other hand, will always have the disease if they possess the gene, and all their daughters will be carriers (since they can only pass on an affected X). All their sons, however, will be completely free from the disease as they will receive only a Y from their affected father. This should be apparent but it is surprising how many people with quite advanced medical knowledge do not appreciate this fact. I often used to ask candidates in medical examinations what proportion of a male haemophiliac's sons will develop the disease provided he marries a normal woman. Nearly all candidates pause, look very wise and then say 'about 50%, sir'! It must be remembered that the *men* in these affected families can always know what their children will be. The normal ones in them will have all normal children by normal wives (and they will never develop the disease later in life as it always manifests itself early), whereas haemophiliac men, if they live long enough to have any children at all, will have all normal sons and all carrier daughters. It is those women who have, say, a haemophiliac brother, and who may have inherited the gene from a carrier mother who are in doubt, and it is sad that they cannot know for certain whether or not they are carriers, though about 85% (see p. 34 for explanation) of these heterozygote carriers can be detected by assessing the degree of factor VIII reduction in their plasma.

Factor VIII is very much in the news at the present time because it can be engineered and therefore should be free of the virus that causes AIDS. However the clinical as opposed to the *in vitro* efficiency of the product is still unknown.

A side issue of the synthesis of the factor VIII molecule is that part of it resembles in its amino acid sequences the molecule of the quite unrelated protein ceruloplasmin which bonds copper atoms and its function in human plasma may be to convert ferrous (Fe^{2+}) into ferric (Fe^{3+}) ions. Perhaps evolution may have hit on a way of using a pre-existing molecular structure (ceruloplasmin) as a method of helping blood to clot rather than carrying out its normal function of banishing Fe^{2+} ions.

Before leaving haemophilia it is perhaps appropriate to give a résumé of how blood coagulation works. The end result must be a plug to stop the bleeding, but the steps involved in this are numerous – constriction of blood vessels, aggregation of platelets to damaged surfaces, and finally the formation of fibrin clots. These are all regulated by enzymes which function as a cascade, each releasing an active factor from an inert precursor and all are under genetic control. Therefore at each step there may be abnormalities and most of these mutants and variants have been recognized from a study of patients, haemophilia, as will be appreciated, being the classic example. But lack of 'contact factors' may also inhibit coagulation, for example a defect in factor XI which produces a bleeding disease found particularly in Ashkenazi Jews. Factor XIII the last in the line is responsible for the stabilization of fibrin and its deficiency (controlled by an autosomal recessive gene) may lead to a faulty clot with the result that bleeding occurs from the umbilical stump a week or so after birth.

2.2   The fragile X

X-linkage is also important in mental retardation because it has recently been shown that there is an inherited form due to a chromosomal abnormality – the fragile X syndrome – so called because in squash preparations (in which the cultured material is 'squashed' between glass slides) the X chromosome readily fractures at a particular site (Turner *et al.*, 1978). The affected boys have a tendency to big ears, a protruding chin, large hands and feet and large testes.

The syndrome is of extreme current interest because it still contains many uncertainties. Thus, occasionally, carrier females are themselves mildly mentally retarded, and more importantly, there is an increasing number of reports of transmission through males who appear perfectly normal. Such males have heterozygous daughters who are never mentally retarded and have either no fragile sites or very few indeed. By contrast, in the subsequent generation a third of the female heterozygotes are mentally subnormal, with an average of 29% fragile sites. This information suggests that there is a premutation which generates the definitive mutation only when transmitted by a female. It is proposed that initially there is a mutation that causes no ill effect *per se* but generates a significant genetic imbalance when involved in a recombination event (in a woman) with the other X chromosome. This hypothesis (Pembrey *et al.*, 1985) explains many of the puzzling aspects of the syndrome, but a great deal more work needs to be done, particularly as the subject is related to the question of therapeutic abortion of carrier women who are pregnant with male fetuses.

2.3   Duchenne muscular dystrophy

This is an appalling disease of young boys which almost always ends fatally before the age of twenty years. It starts with a waddling gait and then increasing difficulty in climbing stairs. The thigh muscles are wasted but there is pseudo-hypertrophy of the calf muscles. By about the age of ten, patients have to use a wheelchair and in their early teens they usually realize that they are going to die – they are mentally perfectly normal) – whereas earlier they have just regarded themselves as the semi-invalid of the family.

Since affected boys hardly ever reproduce (Harper, 1981), the disease is transmitted almost entirely by healthy carrier females. Serum creatinine kinase is the standard test and is usually raised in carriers, but there are many pitfalls. It is also sometimes of help to test for muscle weakness in potential carriers. Very recently (1985) using a gene probing technique (see Chapter 12) much more accurate prenatal diagnosis and carrier detection became possible.

The rules for simple mendelian inheritance may be modified by information in particular families. For example, a woman who has had two affected sons *must* be a carrier. Her daughter *may* or *may not be* but if she had four normal sons the risk for a subsequent pregnancy would be much reduced.

There is a much milder form of the disease, Becker muscular dystrophy, also sex-linked, which has different clinical features and a later onset. Patients often reproduce; all their daughters will be carriers and all their sons normal (see Fig. 2.2 for positions on the X chromosome).

2.4   The mapping of the X chromosome

Crossing-over (see glossary) will be discussed in Chapters 4 and 8, as will the method of calculating the cross-over value, which indicates how far apart the genes are likely to be on the chromosome (p. 41).

Quite a number of genes are known to be situated on the X chromosome (some concerned with disease but many not), the best known among them being those controlling colour vision, two types of muscular dystrophy, the skin disease of ichthyosis, the two forms of haemophilia already mentioned, G6PD deficiency (see p. 71) and the very important Xg blood group system. The last two conditions are detectable in the heterozygote, and the frequency of the Xg heterozygote in the European female population (i.e. the frequency of women having the gene controlling the Xg^a blood group antigen on one X chromosome but not on the other) is as high as 46%. Many families, therefore, which segregate for one of the rare sex-linked conditions mentioned above will also segregate for Xg, and one does not have to search for families segregating for two rare traits. It thus becomes possible to find the relationship of the rare mutant gene to the Xg locus, and thence to one another. The distances are measured in cross-over units, or map units, one map unit being equal to 1% of crossing-over. Thus if 5% of the offspring of informative matings (see p. 41) are of the 'recombinant' type – that is, they are the result of crossing-over – then the two gene loci are said to be about 5 map units apart. Work is in constant progress so that the map is often altered, but Fig. 2.2 is reasonably up to date.

2.5   Sex limitation or sex control

It was mentioned earlier that genes are said to be sex-controlled or sex-limited when they can express themselves only in one sex, and this is quite different from sex-linkage. An instance is described here for comparison. Frontal baldness only shows itself in males except on the rare occasions where a female receives the gene in double dose. The fact that an affected male can transmit the condition to his son shows that the gene responsible cannot be on his X chromosome, and the fact that he can transmit it to his daughter shows that it cannot be on the Y. Thus the gene is on an autosome but can only express itself in the male gene complex – why this is so is unknown.

2.6   Expressivity of genes

What has been written in the first two chapters is simply standard genetics but we need to know more than whether a gene is dominant, recessive or sex-linked. Of equal interest is why genes only express themselves in the appropriate organ and here there are new discoveries.

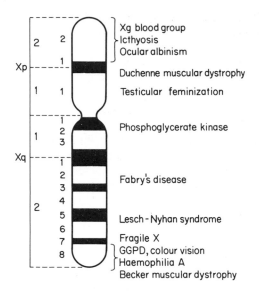

Fig. 2.2 A tentative map of the X-chromosome.
Key to conditions mentioned
Xg blood group — (see p. 12).
Ichthyosis — one form of an inherited condition associated with dry skin.
Ocular albinism — patchy pigmentation of the retina and iris.
Duchenne muscular dystrophy — (see p. 11).
Testicular feminization — apparently normal females but with XY chromosome constitution.
Phosphoglycerate kinase — an enzyme deficiency associated with one form of haemolytic anaemia.
Fabry's disease — a disease of skin and blood vessels.
Lesch-Nyhan syndrome — mental retardation, self mutilation, neurological defects.
Fragile X — (see p. 11).
G6PD, colour vision — (see p. 71).
Haemophilia A — (see p. 8).
Becker muscular dystrophy — (see p. 11).
Key to the recording of the position of the gene on the X-chromosome
Xp = short arm
Xq = long arm
Outer figures refer to regions on the chromosome and are numbered from the centromere outwards.
Inner figures are subdivisions of the regions and refer to the bands (see p. 57)
The position of a gene on a chromosome is always indicated in the same order — first the chromosome number (or X or Y), then whether on the short or the long arm, p or q, then the number of the region and finally the number of the band in that region.
Thus the G6PD locus is at Xq 28
(Adapted from Emery (1983), by courtesy of the author.)

Thus in *Drosophila* it has been found that a dominant mutation could give a fruit fly a pair of legs where antennae should be. This led to the finding that a mutation of the particular gene also affected the expression of others – in what appeared to be a hierarchy lying close together in two clutches on chromosome 6 of the insect. The crucial segment of DNA is called a homoeobox and is a short sequence of 180 base pairs which has a decisive influence on the earliest developmental decisions made by the first embryonic fruit fly. As far as Man is concerned the most important discovery of all is the recognition that the genes in the homoeobox of *Drosophila* have sequences in common with genes of (so far) unknown function in higher organisms, including Man.

3
Multifactorial Inheritance

3.1   Historical background

Mendelism was at first thought to be completely at odds with the continuous variation studies of Galton (1822–1911). He had measured the correlation between relatives of continuously variable characters, for example height, and found that it decreased in a definite ratio according to the closeness of relationship and that after a few generations most of the ancestral genes would be on average those of the general population. Galton and his followers regarded the all-or-none Mendelian characters as being deviations from the normal doomed to early elimination by selection. The Mendelians on the other hand saw their 'mutations' as being the potential starting point for new species and they regarded continuous variation as being evolutionarily irrelevant (Maynard Smith, 1982). The arguments were settled by the work of the population geneticists Fisher, Haldane and Sewell Wright. They showed that continuous variation could be explained by alternative alleles at many loci, each having a small but additive effect on the phenotype. There was thus in essentials no difference between the views of the two opposing factions: compromise is a well recognized British trait.

3.2   More detailed considerations

From what has been said above it is clear that multifactorial inheritance is not dependent on one single gene (or pair of alleles) for which the individual is homozygous or heterozygous. It is not even dependent on genes only, but on the environment as well, and it means what it says, i.e. that many factors are at work. The genes responsible will all be genes of varying but small effect, and with varying dominance, and it is not always easy to sort out which are the inherited components responsible for the condition and which the environmental ones. The term 'polygenic' refers only to the genetic component of a character controlled by many genes. However, nowadays in spite of the prefix 'poly' it is realized that extensive variability can be conferred by alleles at only a few loci.

It is easy to understand that polygenically controlled characters show continuous variation, whereas those which are controlled by single ones show clear-cut, either/or differences. It is quite simple to visualize the fact that many genes control human height – people are not 'tall' or 'short', they vary through all degrees of tallness and shortness with a few very tall and very short people at the two extremes. The vast majority of adults will fall in an intermediate group and if you draw a graph of their height you will obtain a 'normal curve'. There will be a sex difference and there will be racial differences, of course, and there will be the effect of nutrition, but the genes responsible for height are additive (as are those for intelligence). What is not nearly so easy, and gives rise to a great deal of very complicated mathematics, is the fact that diseases are very often multifactorially controlled. After all, one either has, or has not a disease – it is all very well to say that many genes add up to give you diabetes or a duodenal ulcer, but what constitutes the difference between 'disease' and 'no disease'? Here we come to the fact that there is a threshold beyond which the disease manifests itself – if you have enough genes predisposing to the disease, and also the environmental factors favour its expression, then you will have the disease. Here too there are sex differences, some diseases being more prevalent in one sex than in the other, and Carter (1962) did some very interesting work in this connection. He showed that in pyloric stenosis (see p. 90), which is commoner in boys than in girls, when a *girl* (i.e. the wrong sex) has the disease she has more affected relatives than does a boy suffering from it. The female baby is evidently normally more resistant than the male baby, and when she does develop it this is because she has a very high concentration of predisposing genes; this is why more of her relatives will also have this high concentration of genes and show the disease. This seems rather complicated but it serves very clearly to illustrate the threshold effect seen in multifactorial inheritance.

The study of quantitative variation is of course statistical and the mathematics that are involved are not for the amateur, but Roberts and Pembrey (1978) give a very simple introduction to the principles, which are based on the resemblance between relatives. For each step farther away in relationship, the number of genes in common with one's relatives is halved. A parent passes on to a child half his or her chromosomes; they have half their genes in common. Clearly, the child will pass on half the genes derived from that particular parent, so a grandparent and grandchild will have a quarter of their genes in common, and so on. With a rare dominant autosomal gene, it will be clear on a little consideration that half the children of a person in whom it appears will receive it, a quarter of his grandchildren and an eighth of his great-grandchildren. (If the dominant autosomal gene were not rare, it might come in from both sides of the family and the model would not apply.) Half the parents and sibs of an affected person are therefore like the patient and half are unlike, and the likelihood of the other types of relative being affected can be calculated in a similar way. With multifactorial inheritance, however, many genes are combining together to produce the end result. On the *average*, half of the genes are making the sib or parent just like the propositus (the individual through whom the investigation of the disease had been begun), while half of them are making them no more like him than would be an unrelated subject. The difference is that instead of half the sibs or parents being totally like, and half totally unlike, all the parents and all the sibs are tending to be half-like. All the uncles and aunts are tending to be one-quarter-like; all the cousins one-eighth-like. These measures of resemblance are termed *regressions*, and multifactorial inheritance would follow the pattern indicated if there were no dominance and all the genes were intermediate in effect. This, of course, they are not, so dominance and recessiveness have to be considered and this involves complications. One effect of dominance in multifactorial inheritance, however, remains the same even though there are variations in dominance from gene pair to gene pair, and differences in the amount of effect produced by different genes: *the reduction of the regression for sibs is always half the reduction for parents* (meaning, as will be realized, that sibs are more like the propositus than are parents).

The matter is a complicated one, but the above remarks will give the reader an idea of the arguments involved. The section on high blood pressure which follows describes some of the investigations which are carried out when it is desired to determine whether a condition is multifactorially inherited or controlled by a single gene. It is a dated argument but the principles are still sound.

3.3   The controversy over the method of inheritance of high blood pressure (essential hypertension)

The arterial blood pressure consists of two components, the systolic, associated with the beat of the ventricles of the heart, and the diastolic, which is the constant pressure in the arteries between beats. The upper limit of normal in adults is about 140 mm Hg for the systolic and 90 mm Hg for the diastolic. Occasionally a definite cause, e.g. kidney disease, can be found to explain the hypertension but usually this is not the case and the condition is then said to be 'essential' or cryptogenic. Where this is so the patient often feels perfectly well, the finding being brought to light during some routine medical examination such as for life insurance. However, a family history of hypertension is often found in people with this condition and there is general agreement that inherited factors play a part in its causation. The controversy, however, has been *how* it is inherited, particularly whether as a graded character like height or as a specific 'either/or' disease entity. Some of the evidence which has been brought forward by both sides will now be discussed. The arguments are also relevant to the part played by inheritance in several other common diseases.

Family studies

In 1954 Hamilton and his colleagues investigated the frequency distribution of blood pressure at different ages and concluded that there was no natural division between the normal and the abnormal. They also surveyed the first degree relatives (fathers, mothers or children) of patients with hypertension and found no indication of a bimodal distribution such as would be expected if the population contained two groups, one hypertensive and the other not.

Platt (1959), however, felt that this unimodal distribution in the relatives might for various reasons be masking two groups. He thought it would be of interest to plot the blood pressures of relatives of propositi in a middle-aged group, which is when essential hypertension occurs. By concentrating on this age he largely excluded patients with hypertension due to other diseases.

He therefore took 252 individuals who were all between the ages of 45 and 60, and who were the sibs of propositi also aged from 45 to 60. He plotted the systolic and diastolic pressures of these sibs *only* (leaving out the propositi, all of whom were hypertensive) and found that the blood pressures appeared to fall into two groups, even allowing for some inaccuracies of measurement. Figure 3.1 shows the distribution curves of the systolic pressure of these siblings.

This situation would be consistent with the control of essential hypertension by a single dominant gene, and this is what Platt believes often to be the case.

Population surveys

On the other hand, in 1955 and 1958 Miall and Oldham had carried out two large population surveys. They showed that there was a relationship between the arterial pressure of propositi and their close relatives of the same degree at all ages no matter what the blood pressure of the propositus was. This degree

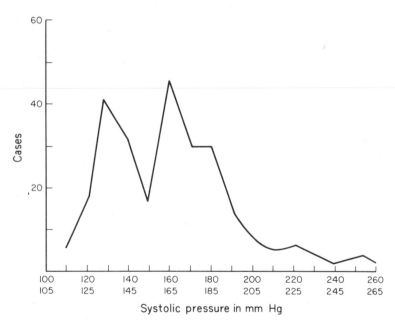

Fig. 3.1   Systolic pressures of 252 siblings aged 45–60 of hypertensives aged 45–60, showing a bimodal distribution curve. The lower row of figures represents a correction to avoid recording blood pressure in even tens. Platt (1959), from data of Hamilton *et al.* (1954), and Søbye (1948). Similar findings are also found with the diastolic pressure. (By courtesy of the authors and the editor of the *Lancet*.)

Table 3.1 Part of a population sample (propositi and first degree relatives) taken from a Welsh survey (Miall and Oldham, 1958). The mean arterial pressures are given in 5-year age groups

Age	Females						Males					
	Population sample (propositi)			First degree relatives			Population sample (propositi)			First degree relatives		
	No.	Sys-tolic	Dias-tolic	No.	Sys-tolic	Dias-tolic	No.	Sys-tolic	Dias-tolic	No.	Sys-tolic	Dias-tolic
5—	11	106.6	70.2	52	105.3	69.0	12	105.4	72.1	38	105.3	68.4
10—	10	112.0	72.5	47	112.5	71.8	9	110.8	73.6	40	114.6	73.3
15—	10	125.0	80.5	42	119.2	72.8	18	128.6	78.3	33	120.8	75.2
20—	10	122.5	75.0	43	121.2	74.8	6	126.7	78.3	37	129.5	80.9
25—	2	112.5	62.5	41	123.5	77.2	17	128.4	82.5	29	123.4	78.5
30—	13	121.4	74.8	52	126.3	80.5	12	131.3	84.6	45	129.3	84.2
35—	14	130.7	82.9	46	129.9	82.6	11	130.2	85.2	47	125.9	81.4
40—	5	134.5	86.5	47	130.3	82.6	6	128.3	85.0	33	126.6	82.7
45—	5	148.5	88.5	38	137.6	84.9	9	134.2	83.1	40	127.9	80.8
50—	8	149.4	88.8	30	148.8	87.8	7	137.5	86.1	32	141.6	89.4
55—	8	169.4	98.8	30	160.8	89.5	8	150.0	88.8	18	148.1	89.7
60—	8	180.0	93.1	19	170.1	93.8	8	146.9	87.5	25	144.1	84.7
65—	4	190.0	97.5	17	169.9	91.9	6	157.5	88.3	18	167.8	92.8
70—	3	200.8	102.5	10	173.0	91.0	3	154.2	94.2	7	148.9	85.4
75—80	2	210.0	105.0	7	212.5	103.9	2	162.5	90.0	11	154.3	78.4

Adapted from Miall and Oldham, 1958, by kind permission of the authors and the editor of *Clinical Science*.

(the regression of the blood pressure of relatives on that of propositi) was about 0.2. This means that the relatives of a man with a systolic blood pressure 25 mm Hg higher than the mean for his age, for example, would have a pressure averaging 5 mm Hg above the mean for their ages. Table 3.1 gives some of the actual figures which Miall and Oldham found and Fig. 3.2 shows the data represented graphically, the legend explaining in detail how the regression lines have been constructed. The consistency of the regressions and the normal distribution curve (an example of which is shown in Fig. 3.3) in this large-scale investigation provide obvious support for the hypothesis of multifactorial.

Pickering (1959), summing up the evidence, felt that the multifactorial hypothesis was more satisfying than that based on a single dominant gene. He pointed out that arterial pressure, like height, was the result of a large number of variables, the elasticity of the vessels, the radii of different parts of the vascular system and the action of the heart all playing a part. Platt (1959), on the other hand, argued that only one of the many variables may be disturbed in any particular group of hypertensives, and gives as an example the condition of aldosteronism, which is a specific endocrine upset causing hypertension, the successful treatment of which takes the patient out of the hypertensive group.

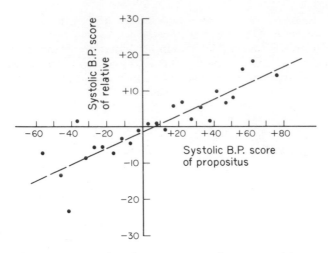

Fig. 3.2 Systolic blood pressure scores of propositi and the mean systolic scores of first degree relatives. The scores represent the deviations of blood pressure from the mean, higher pressures than the mean having positive scores and lower pressures negative ones. Each dot represents a pair of measurements. It will be seen that propositi who, for example, are + about 80 on the mean have first degree relatives whose mean pressures are about + 17. The scores have been adjusted to take age and sex into consideration. The regression for the diastolic blood pressure was similar. (Miall and Oldham. Reprinted by permission from *Clinical Science*, vol. 17, pp. 409–44, copyright © 1958. The Biochemical Society, London.)

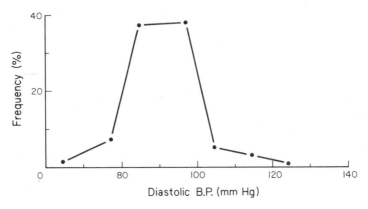

Fig. 3.3 Frequency distribution curve for 84 sibs aged 45–59 derived from propositi aged 45–59 with diastolic pressures of 100 mm Hg or above. Adapted from Oldham *et al.* (1960). *Lancet* **i**, 1085–93. (By courtesy of the authors and the editor of the *Lancet*.)

Twin studies

The final paper of interest is that of Platt who in 1963 investigated the blood pressure in monozygotic identical twins, and in three pairs where the propositus had severe hypertension so had the co-twin, the measurements being:

Propositus	*Twin*
260/150	210/130
230/130	205/130
200/130	210/130

This concordance argues in favour of inheritance rather than environment, but it obviously does not tell you whether the method of inheritance is multifactorial or single gene. On the other hand. Platt's data with regard to non-identical (dizygous) twins, where one of the twins or a non-twin sib had hypertension, suggests a single gene situation – hypertensive or normal. We are impressed with the twin studies, though the numbers are small, but remain unconvinced by his view that essential hypertension is always a unifactorial inherited disorder of middle age.

It may well be that the conflicting views of Pickering and Platt are not irreconcilable. After all, height is controlled multifactorially and may be secondarily affected by superimposed genetic or environmental factors, for example, various forms of dwarfism on the one hand and malnutrition on the other. Whatever be the truth of the matter a good argument often teaches one more than the straightforward 'establishment' version of a disorder, and medical conferences were enlivened by speakers 'bickering with Pickering or having a bat at Platt'. (For a much later reference see Swales, 1985.) We may not yet have heard the end of the story, for with the much more efficient treatment of high blood pressure, it is interesting to see the variability of response by different individuals to different drugs (see Chapter 9).

Possible site of the lesion in essential hypertension

Curtis *et al.* (1983) report on the remission of essential hypertension in six patients after renal transplantation. All the sufferers had end stage renal disease and in five of the six the hypertension persisted after removal of their kidneys, but before transplantation, while they were on dialysis. After receiving transplants from normotensive donors the patients themselves became normotensive and have remained so for from one to eight years, with normal renal function. These findings are clearly very suggestive of a kidney lesion being the primary cause of essential hypertension perhaps related to the transport of sodium. Salt has been regarded as a possible environmental factor because there is a correlation between salt intake and blood pressure both between and within racial groups.

4

Genetic Polymorphism

4.1   Definition and general considerations

Genetic polymorphism is a type of variation in which individuals with clearly distinct qualities exist together in a freely interbreeding single population. Ford (1940) defined the condition as 'the occurrence together in the same habitat of two or more discontinuous forms or "phases" of a species in such proportions that the rarest of them cannot be maintained merely by recurrent mutation'. This definition excludes several familiar types of variation. For example, the Caucasian, Mongolian and Negroid races of Man do not constitute a polymorphism since when interbreeding occurs, the hybrid populations are intermediate and variable. Again, continuous variation, as in human height, is not an example of polymorphism. In these examples, many genes are at work and the variation is brought about by the cumulative effects of segregation taking place at many loci, and not by 'switch' genes giving rise to distinct alternative forms. Seasonal forms, too, are excluded from the definition. For instance, in the Map butterfly, *Araschnia levana*, temperature or length of daylight can produce very distinct spring and summer forms but in this type of situation all members of a generation are alike and this does not constitute a polymorphic system. In addition, segregation in human populations into normals and phenylketonurics or normals and achondroplasics does not fall within the definition since these diseases are constantly being eliminated by selection and are maintained by recurrent mutation.

The polymorphic type of discontinuous variation which we are discussing is nearly always genetic, and in a polymorphic system a continuous range of intermediates is absent. There must therefore be some very accurate switch mechanism controlled either by alleles at a single locus (such as those determining the ABO blood groups) or by the corresponding members of a supergene (see p. 24) which produces either one form or another.

4.2   The establishment and maintenance of polymorphism

How then does a polymorphism arise, and how is it maintained? It arises initially by mutation, and the selective effect of a mutant must be either disadvantageous (which it usually is), neutral or advantageous. In the first case, the mutant will never be anything but rare, as it will consistently be selected against. With respect to the neutral states Fisher calculated that the balance of a mutant and its normal allele would have to be extraordinarily exact for the two to be neutral in effect. This is because mutants responsible for very trivial visible effects e.g. eye colour and bristle number, in fact alter the length of life of the insect and its capacity to survive under unfavourable conditions. Furthermore, in the exceptional event of a mutation being neutral its spread would be exceedingly slow. It follows, therefore, that where a polymorphism exists the third situation must have been operative, and the mutant gene compared with the other alleles must, under certain circumstances, have been at an advantage. If, however, the advantages were complete in every respect, the mutant gene would simply be on the way to replacing the original one and the polymorphic situation observed would be a temporary phenomenon ('transient' polymorphism). The polymorphisms with which we are concerned in this chapter are, so far as is known, not transient. They are 'balanced' or 'stable' polymorphisms, arising (in a manner described later) because for some reason discontinuous diversity is advantageous (see Fisher, 1930).

4.3 Heterozygous advantage and the evolution of dominance

The usual way in which polymorphisms are maintained is through the selective advantage of the heterozygotes over both homozygotes – this will keep the alternative alleles in the population. This heterozygous advantage can arise in two ways. First, as Ford (1965) very clearly explains, any gene which begins to be favoured must exist almost entirely in the heterozygous state in the initial stages of its increase since there is little chance of matings between the rare heterozygotes. If the mutant gene has a slight advantage, recessive lethals or semi-lethals which are carried on the same chromosome can be sheltered from elimination by their proximity to the advantageous mutant, *as long as they are in a heterozygous individual*. As the newly successful supergene increases in frequency, homozygotes appear and will be handicapped by these harmful recessives (now homozygous and consequently active). Secondly, major genes always appear to have multiple effects and if one of the features for which a mutant is responsible gives it an advantage and others do not, selection will tend to make the beneficial effect dominant and the harmful ones recessive (Sheppard, 1975). In these circumstances the homozygotes will have both advantages *and* disadvantages since they will be homozygous for the deleterious recessive genes as well as for the successful mutant, while the heterozygotes will bear the advantageous dominant allele but only one dose of the disadvantageous recessive ones.

(The terms 'recessive' and 'dominant' should strictly never be used of genes, only of the characters they determine, since a single gene can have many effects and the same gene will often control both recessive and dominant characters.)

The above is the classical view of genetic polymorphism but in recent years gel electrophoresis has demonstrated that a large amount of enzyme variation is polymorphic and the view has been put forward that, in spite of Fisher, the genes responsible have neutral survival value and the polymorphism *is* maintained by mutation. Kimura is the key name associated with neutral genes. Ford (1973) has put forward objections to the views of Kimura and others but the fact remains that in the vast number of enzyme polymorphisms which have been described the nature of the heterozygous advantage remains unknown.

4.4   The formation of a supergene

Particularly relevant to the establishment of polymorphisms is the formation of supergenes and this will be discussed in relation to certain forms of mimicry in butterflies, where the details have been clarified and can be readily understood. Here, within a single species, there are various forms of female each of which obtains a selective advantage by resembling another species which is distasteful to predators, such as birds. A polymorphism therefore arises and this remains balanced because an excess of mimics resembling one distasteful model would result in the predators beginning to associate that particular wing pattern with edibility, and not with inedibility. However, the mimetic wing pattern is complicated, and experiments have shown that although the 'gene' controlling it behaves as a single unit it is in fact composed of separate genes which have come to lie close together from different parts of the same chromosome or even from non-homologous chromosomes. Occasionally crossing-over occurs, breaking up the advantageous combination of the supergene so that unusual patterns may be seen which will in general be selected against since the mimicry will then be less good.

   A similar situation may have arisen in Man in the Rh blood groups. Here, there appear to be three loci controlling the antigens C or c, D or d and E or e. The corresponding antibodies are anti-C, anti-c, anti-D, anti-d (so far hypothetical), anti-E and anti-e. The commonest Rh combinations on one chromosome are CDe, cde and CDE, and it is postulated that the rarest ones (e.g. CdE) have arisen as the result of crossing-over. However, the matter is much more complicated than this, for there are in fact about 20 Rh antigens and antibodies and there are probably very many mutational sites at the various loci. Furthermore there has been shown to be a 'combined' antigen depending on whether c and e are on the same chromosome or on the homologous one. Testing for this antigen has demonstrated that the c and e genes are in the same cistron (a microbiologist's word which approximates to the term supergene). The cistron is the portion of the chromosome in which the loci are integrated for one function. The argument that c and e are in the same cistron is as follows. When genes are on the same chromosome (that is, inherited from the same parent) they are said to be in *cis*, and when the c and e genes are in *cis* they produce ce antigen (formerly called f). The c and e genes are therefore *non-complementary* – they cannot combine to form the product when they are on different chromosomes though they can when they are on the same one. According to current definition, this means that they are in the same cistron. If they *could* combine when in *trans* to form ce, they would be complementary and this would mean that they were in different cistrons.

Though it seems reasonable to regard the Rh system as forming a super-gene, why it has come about is unknown, but the CDE combinations may have varying selective advantages in different genetic constitutions and it is known that the antigens have widely differing antigenicities.

4.5 Chromosomal inversions giving rise to chromosomal polymorphism

Another example of the way in which a collection of genes can operate as a unit is when there is a chromosomal inversion giving rise to chromosomal polymorphism. It has been demonstrated in *Drosophila pseudo-obscura* and *D. persimilis* that chromosome polymorphism can be maintained by heterozygous advantage (heterosis). Numerous instances have been found in wild populations in which the numbers of inversion heterozygotes exceed expectation, assuming equal viability for all three genotypes. An entirely independent proof of heterosis is provided by the fact that when a population of larvae carrying the inversions is reared in the laboratory under optimal conditions, the three genotypes in the adults have the normal proportions expected from the Hardy-Weinberg law (see p. 31). When, however, the larvae are in competition for a restricted food supply it is found that the proportion of heterozygotes exceeds expectation among the adults to which they give rise.

In some cases the polymorphism consists in the number of chromosomes. For example in *Nicandra physaloides*, distantly related to the tobacco plant, the fertilized seeds with one chromosome fewer than normal are subject to delayed germination and the variation so produced is of advantage in contending with environmental fluctuations.

4.6 Some polymorphic systems in Man

A number of polymorphic systems are mentioned in other chapters of this book but a few of particular interest are described here in more detail.

The sickle-cell trait

In spite of the large number of polymorphic systems which have been described in Man, the selective factors involved are known in only very few. The classical example is that of sickle-cell and normal haemoglobin, the genes controlling which are situated on one of the autosomes. That controlling the formation of haemoglobin S produces the sickle-cell *trait* when accompanied by the normal allele and sickle-cell *anaemia* when in double dose. The Hb^S gene has a frequency exceeding 20% in several East African populations and this means that about 4% of new-born children are homozygous Hb^S/Hb^S and almost all of these will die in infancy. The reason for the high frequency of Hb^S is that the heterozygotes have a considerable advantage over normal people (Hb^A/Hb^A) because when young they are more resistant to infection by malignant tertian malaria (Allison, 1954) and they obviously have an advantage over those who are Hb^S/Hb^S.

The ABO blood group polymorphism

The known world distribution of the blood group frequencies is strong evidence for the action of natural selection on some at least of the blood group systems. Although the relative frequencies of the phenotypes A, B and O vary markedly over very short distances, some of the other systems show much more gradual change with distance. This may be, in part, due to lack of detailed knowledge, but there is no doubt that there is a real contrast in the geographical variation between the ABO groups and some of the other systems. Therefore, as E. B. Ford suggested, it was well while looking at diseases to see if a particular ABO blood group predisposed (even though very slightly) to particular disorders. What follows shows that this was so, and it is interesting to note that ABO compatibility is important in transplantation (cf. the HLA system and associations with disease, see p. 53).

Differential susceptibility to infectious diseases

One good reason for particular attention being paid to some infectious diseases is the fact that several micro-organisms are known to possess antigens very similar to human blood substances. The thesis is that the world distribution of the ABO blood groups may have been influenced by the history of the great pandemics of infectious diseases of former times, particularly those with a high mortality such as plague and smallpox. The most interesting studies were carried out in relation to smallpox, as in this case there was the opportunity to study the problem in those parts of Asia where smallpox was highly endemic. It was predicted that smallpox should show a more severe course and a higher mortality in those individuals of blood groups A and AB than in those of groups B and O. The argument was based on experimental evidence that the vaccinia virus, and by inference the variola (smallpox) virus, possessed an antigen similar to blood group A substance. Thus, blood group B and O individuals who possess a natural anti-A antibody might be expected to be more likely to neutralize the virus during the viraemic stage and so sustain a milder course of disease. The hypothesis received adverse criticism and evidence was produced that the A-like antigen demonstrated was present *in the egg material on which the vaccinia virus was grown and not in the virus.*

The experimental basis for the thesis in relation to smallpox was thus in considerable doubt. However, the epidemiological evidence is of great interest. It was found that reactions to vaccination and, in particular, encephalitis (inflammation of the brain) were more common in people with the blood group A gene than in those without. Several studies of the natural history of smallpox in patients of different ABO blood groups were made with conflicting results. A survey was therefore designed to test the hypothesis under more critical conditions and to avoid some of the faults of previous studies. The research workers aimed at the complete ascertainment of cases of smallpox occurring over a certain period in selected highly endemic areas of West Bengal and Bihar (India). The survey was confined to rural areas where few people had been previously vaccinated and where little modern medical treatment was available. The choice of a control population for such a study is always a matter of some complexity. In populations as mixed as the ones studied there is always the possibility that one section of the population was inbred for a long period, leading to a different distribution of ABO genes from that of the rest of the population, and that this section might for *other* reasons have a greater or lesser susceptibility to smallpox. In this way an apparent association between blood group and susceptibility might be found (stratification effect). To avoid this the severely exposed but unaffected siblings were used as controls.

It was shown that individuals of blood groups A or AB *were* much more liable to develop smallpox than those of groups B and O and, moreover, the course of the disease tended to be more severe in the former. The mortality was significantly higher in individuals possessing the blood group A gene than in those without.

The above is now only of academic interest since vaccination (discovered 200 years ago) has, at last, put an end to the disease which will not reappear (except perhaps from corpses) as there is no animal carrier.

It is a good example of medical advance outwitting one's inheritance – and a triumph for preventive medicine.

Differential susceptibility to chronic 'adult' diseases

As is discussed in Chapter 7 duodenal ulcer is commoner in people who are group O than in those of the other ABO groups. Cancer of the stomach is more frequently found in individuals who are group A and a similar finding is also reported in pernicious anaemia, a disease in which there is no acid secretion in the stomach (as is often the case, too, in stomach cancer). However the effect of these diseases on the frequencies of the A, B and O genes is certainly very small since the mortality from them chiefly occurs after the reproductive period is over – but this may not always have been so.

In any case, the findings were of little or no medical importance, and since the last edition of this book duodenal ulcer and cancer of the stomach have become much rarer whereas the blood group proportions in the population remain unchanged. The ABO group associations have been superseded by the HLA associations but how far these actually help patients with a particular disease raises the same sort of doubts as did the ABO group work 30 years ago.

The Rh blood group polymorphism

Here we have a most puzzling problem. Why, when the most important mechanism for keeping a polymorphism in existence appears to be the advantage of the heterozygote, do we find in the Rh situation that there is strong selection against the heterozygote? Any fetus born of an Rh negative mother, if it be Rh positive, must be heterozygous and is in danger of suffering from Rhesus haemolytic disease (see p. 81) and though this is by no means invariable, the fact that the loss is always in the heterozygote constitutes a serious difficulty in explaining the polymorphism. Even if the homozygotes were equally advantageous, the fact that for every child that dies there is destruction of one D and one d gene would mean that whichever of the two alleles was rare to begin with would eventually be eliminated.

Several explanations have been put forward to explain why this has not taken place. The first is that parents who have lost a child will 'compensate' to replace it, and end up by having more children than those parents who had lost none. However, this cannot be the whole explanation because it has been shown mathematically that compensation will not lead to a polymorphism which is stable. Thus below a certain critical gene frequency, the value of which depends on the compensation effect, Rh negatives will still tend to decrease in frequency but above it they will tend completely to replace Rh positives in the population. Consequently we need some other selective effect to account for the persistence of the polymorphism. There is the possibility that in very primitive populations a reduction in the number of children owing to deaths from haemolytic disease would leave more food for the rest, thereby increasing the mean number of children reaching maturity in small families above that in large families, under famine conditions. This has been known to occur in birds, where the parents were unable to feed large clutches adequately. However, it seems a very speculative explanation in Man.

Another suggestion is that the polymorphism may have occurred in the first place by the d gene establishing itself and increasing owing to genetic drift (see glossary) and being thereafter selected for – why we do not know – until there were two races, one high in D and one high in d. Race mixture after that would account for the high d frequency in Western Europe, support for this being that there are even now high d races – among the Basques and the Berbers. Several originally high d races have also been thought to be a possibility, but if this were the case it is surprising that there is not more variation in the frequency of the gene in different areas. (See Penrose, 1959, for discussion on natural selection in Man.)

The histocompatibility leucocyte antigen (HLA) system

The most complex example of a supergene in Man is the HLA system, which is so important in organ (e.g. kidney) and tissue (skin) transplantation, and in its associations with an increasing number of diseases.

The system was discovered via its antibodies, which arise in various circumstances. For example they are made by patients receiving multiple transfusions, individuals deliberately immunized by skin grafts or leucocyte injections, patients who have rejected an organ graft and women who have had multiple pregnancies and who have become immunized by fetal antigens derived from the father; it is interesting that here the antibodies do not usually harm the baby.

Sera from such individuals were tested for lymphocyte agglutination reactions against the white cells of many randomly selected donors. The results obtained with each serum were compared with those obtained with every other serum (by computer analysis) and this enabled the sera to be classified into a relatively small number of groups. This approach, together with the study of families, have defined the five HLA loci, all on chromosome 6, and called A, B, C, D and DR. At each of these loci there are many (numbered) alleles, so that the numbers of combinations are very great, and before transplantation the best possible matching has to be assured.

This is probably less necessary than it was a few years ago since the antibiotic cyclosporin is very effective in combating rejection reactions. It seems to disable only those immune cells which the body has produced to reject a particular transplant.

A most interesting phenomenon is the effect of a previous blood transfusion on the recipient. Previously, an earlier transfusion was thought to be a strong contra-indication to a transplant because of the antibodies which would have been formed. However, most experts are now of the reverse opinion. The precise reason for the protective action of a transfusion is unknown, but it may be the result of some feedback system which induces the patient to manufacture cells which suppress rejection.

Most important of all in transplantation is a good surgeon and a fresh kidney – and ABO compatibility between donor and recipient is a 'must'.

Associations between certain HLA types and disease are mentioned in Chapter 7, section 9, on p. 53.

4.7 Infinite variability

Antigens and antibodies have been discussed in what might be thought of as their developed state, but until recently there remained a most perplexing biological problem – how animals could produce (as they do) antibodies against *any* antigen, of which there is a vast array. An answer has now been found, greatly aided by work on the mouse, which has clarified the structure of the immunoglobulin (antibody) molecule. This is basically simple, the molecule consisting of two identical heavy and two identical light chains (Fig. 4.1) all made up of sequences of amino acids, their exact structure being determined by the DNA in the genes, as in any other polypeptide.

There are, however, special and very complex genetic arrangements at work. Thus, essentially, the molecule consists of certain parts which are highly variable and some which are constant. This, together with the fact that mutation is commoner in lymphocytes than in other cells, and that crossing-over is frequent, gives enough diversity of antibody structure so that when *any* foreign antigen is encountered there is already present an antibody which will recognize and fit it. Processes are next set in motion to produce multiple copies

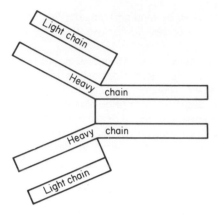

Fig. 4.1   The basic structure of an immunoglobulin.

of that particular antibody, and the individual is then said to be immunized against that antigen.

One practical application of this mechanism concerns the diagnosis of cancer of the lymphoid tissues. The normal population of B-cells (glossary) can be extensively rearranged in many different ways, and, as would be expected, if their DNA is subjected to restriction enzyme digestion (see Chapter 12) myriads of fragments of DNA of widely varying sizes are obtained with no single size predominating. On the other hand, when there is a cancer of the B-cells this will have arisen from a single cell and so the population becomes 'monoclonal', each cell possessing the same DNA (see Chapter 12). In this way, cancer of B-lymphocytes can be differentiated from benign proliferation of lymphoid tissue, which often occurs in inflammatory processes.

All this may seem a far cry from genetic polymorphism, but the either/or examples of antigen/antibody complexes cited earlier in this chapter must be the result of balancing advantages and disadvantages, evolved over millions of years. However, this is not the usual sequel, for example, there is no disadvantage in having an antibody to measles.

Perhaps we should end on a philosophical note. Animals reject grafts because they have an immune system, whereas plants, lacking it, accept them; but how did the divergence begin, and why is it that the two great categories of living things can each get along so nicely, the one possessing and the other lacking a system which might have been thought to be a prerequisite for success?

5
Genes in Populations

5.1 The explanation of the Hardy-Weinberg law

This is a theoretical chapter and in previous editions it was written 'It is not an overstatement to say that when a student of medicine really grasps the principles underlying it he or she has made a very great step towards understanding genetics'. While this remains true yet another aspect is the ability to comprehend the difficulties or facility of the law. For example, part 1 states that, *with certain provisos*, if there are two or more contrasting alleles (e.g. those controlling the ABO blood groups) in a population their proportions will remain the same from generation to generation. The provisos are that neither allele is being selected for, or against, and that random mating is taking place − that is, any individual has an equal chance of mating with any other individual with regard to the trait in question. The difficulty here is to appreciate how simple this all is. If there are no alterations in the environment causing selection and random mating is taking place why *should* the proportions change, surely they will go on happily generation after generation? However, other people think differently, querying, for example, why dominant traits do not increase at the expense of recessive ones. After all, on average, three-quarters of the progeny of two heterozygotes will manifest the dominant trait but only one-quarter the recessive one. There will, therefore, be three times as many progeny manifesting the dominant as the recessive trait − *but* the point of the law lies in the proviso that mating in the population is *random* and therefore the proportion of the recessives does *not* diminish.

Part 1 of the law is of most practical interest when it demonstrates that a population is *not* obeying the provisos; for then it can be inferred that either natural selection or non-random mating is taking place. For example, an increase in the mutation rate, genetic drift (see glossary) and migration could all lead to changes in gene frequencies as could selective factors. Taking a specific example from biology, consider the frequency of the pale form of the peppered moth compared with the black one. If random mating is the rule (as it is) one can see that the Hardy-Weinberg equilibrium is not maintained and the frequency of the typical forms go up year by year. We therefore look to natural selection and find that the reason for the increases in the pale form are the result of the implementation of the Clean Air Acts − pollution has decreased and the black form is less well camouflaged.

For medical personnel there are only a few reasons for finding out gene rather than phenotype frequencies. Thus, the experts point out that the former allow theories concerning the manner of inheritance of a gene to be tested, and knowing the gene frequencies, the expected frequency of children of different groups from any type of mating can be calculated. Also, in ABO blood typing, if in a given sample the Hardy-Weinberg equilibrium is in operation, one can then have confidence in the antibodies used for defining each group.

Table 5.1 reminds us that heterozygotes (and often, therefore, carriers) are much commoner than would at first glance be expected.

5.2 Algebraical considerations

Dr D. A. Price Evans has set out in detail the mathematical aspects of the equilibrium.

First, it is necessary to understand that a group of individuals can be represented by a 'gene pool'. Say, for example, there were 100 people in a room, one could if one was considering the ABO blood group system stop thinking of them as individuals and just think of the ABO blood group genes which they represented. Each individual has two blood group genes, so the room would contain 200 ABO blood group genes. This is a 'gene pool' and within it there would be some A, some B, and some O allelic genes.

Using a system with three alleles as an example makes the explanation of the next step too complicated, so we will leave the ABO blood groups now that the idea of a gene pool has been described.

Supposing that there are two hypothetical alleles A and a which can occupy an autosomal locus. In the gene pool the proportion of A as a fraction of 1 is p. This could be say 0.7 (i.e. 70%). Then the fraction of the pool due to allele a would be 0.3 (i.e. 30%). Unity represents the whole pool.

The genotype of an individual represents two of these alleles in combination. If the combinations occur at random they will have the following frequencies:

Table 5.1

Disorder controlled by recessive gene	Frequency of affected (q^2)	Frequency of carriers ($2pq$)
Fibrocystic disease	About 1 in 2000*	1 in 22
Albinism	About 1 in 20 000 (0.00005)	1 in 71.9 (0.0139)
Phenylketonuria (see p. 5)	About 1 in 25 000 (0.00004)	1 in 80 (0.0125)
Amaurotic family idiocy (a lethal condition associated with blindness)	About 1 in 40 000 (0.000025)	1 in 100.5 (0.00995)
Alkaptonuria (a rare inborn error of metabolism associated with darkening of the urine on standing)	About 1 in 1 000 000 (0.000001)	1 in 502.5 (0.00199)

* On the assumption that it is not heterogeneous.

$$\begin{array}{ll} \text{AA} & p^2 \\ \text{Aa} & 2pq \\ \text{aa} & q^2 \end{array}$$

Now supposing these individuals mate at random, what is the result? It can be found by first painstakingly writing down all the possible genotypic matings which can occur and this is shown in the left hand column of Table 5.2. In the next column, the frequencies are given.

The next question is what sort of offspring will arise from each mating? This can be worked out very simply by considering the genotypes of the parents, and the frequencies at which the different offspring occur can be entered in the appropriate columns. This is shown in the right hand part of Table 5.2. Now the algebraically minded can apply themselves to each column of offspring, and find that the terms in the column headed AA reduce to p^2, the terms in the column Aa reduce to $2pq$ and the terms in the column aa reduce to q^2.

In other words, what comes out in the next generation after random mating is a population with the same frequencies as before. The size of the population might be larger or smaller but the structure would be the same. That is unless something happened to upset the equilibrium. The factors which can do this are assortative (i.e. non-random) mating, mutation, selection and genetic drift.

Two more points about Table 5.2 need considering. You will see that the mating of heterozygotes gives offspring in the 1:2:1 ratio of genotypes. When there is dominance i.e. Aa resembles AA then proportions of dominants to the recessives (aa) can easily be constructed from the right hand side of the table.

Offspring column headed aa

$$q^4 + 2pq^3 + p^2q^2$$

Divide through by $q^2 = q^2(q^2 + 2pq + p^2)$

but $\qquad p + q = 1$

Table 5.2

Types of genotypic matings Male × Female	Frequencies of genotypic matings Male × Female	Frequencies of offspring of different genotypes		
		aa	Aa	AA
aa × aa	$q^2 \times q^2 = q^4$	q^4	—	—
aa × Aa	$q^2 \times 2pq = 2pq^3$	pq^3	pq^3	—
aa × AA	$q^2 \times p^2 = p^2q^2$	—	p^2q^2	—
Aa × aa	$2pq \times q^2 = 2pq^3$	pq^3	pq^3	—
Aa × Aa	$2pq \times 2pq = 4p^2q^2$	p^2q^2	$2p^2q^2$	p^2q^2
Aa × AA	$2pq \times p^2 = 2p^3q$	—	p^3q	p^3q
AA × aa	$p^2 \times q^2 = p^2q^2$	—	p^2q^2	—
AA × Aa	$p^2 \times 2pq = 2p^3q$	—	p^3q	p^3q
AA × AA	$p^2 \times p^2 = p^4$	—	—	p^4

so $\qquad (p + q)^2 = 1$

So the sum of the terms in the column $= q^2$

Offspring column headed Aa

$$2pq^3 + 4p^4q^2 + 2p^3q$$

Divide through by $2pq = q^2 + 2pq + p^2$

but, again $\quad q^2 + 2pq + p^2 = 1$

So the sum of the terms in the column $= 2pq$

Offspring column headed AA

$$p^2q^2 + 2p^3q + p^4$$

Divide through by $p^2 = q^2 + 2pq + p^2$

but, again $\quad q^2 + 2pq + p^2 = 1$

So the sum of the terms in the column $= p^2$

5.3   Sex (X)-linkage

When one comes to consider sex-linked conditions like haemophilia then the situation is different because males are hemizygous. So here the females will be just the same as for autosomally controlled conditions with frequencies:

AA	p^2
Aa	$2pq$
aa	q^2

The males will, however, be of two types:

A	p
a	q

The same general plan is followed again in Table 5.3. The types of genotypic matings are listed and frequencies are ascribed to them. Then the types of male offspring produced by each mating have frequencies allotted to them; and finally female offspring are considered separately. Again, the algebraically minded will see that the sum of the terms in each of the five offspring columns gives the same value as the frequency for that particular genotype as in the parental population.

Table 5.3 Sex (x)-Linkage and the equilibrium

Types of genotypic matings	Frequencies of genotypic matings	Frequencies of offspring of different sexes and genotypes				
Male × Female	Male × Female	Male		Female		
		a	A	aa	Aa	AA
a × aa	$q \times q^2 = q^3$	q^3	—	q^3	—	—
a × Aa	$q \times 2pq = 2pq^2$	pq^2	pq^2	pq^2	pq^2	—
a × AA	$q \times p^2q$	—	p^2q	—	p^2q	—
A × aa	$p \times q^2 = pq^2$	pq^2	—	—	pq^2	—
A × Aa	$p \times 2pq = 2p^2q$	p^2q	p^2q	—	p^2q	p^2q
A × AA	$p \times p^2 = p^3$	—	p^3	—	—	p^3

Male offspring

Column headed a

$$q^3 + 2pq^2 + p^2q$$

Divide all through by $q = q^2 + 2pq + p^2$

but $q^2 + 2pq + p^2 = 1$

Therefore the sum of the terms in the columns = q

Column headed A

$$pq^2 + 2p^2q + p^3$$

Divide all through by $p = q^2 + 2pq + p^2$

but $q^2 + 2pq + p^2 = 1$

Therefore the sum of the terms in the columns = p

Female offspring

Column headed aa

$$q^3 + pq^2$$

Divide all through by $q^2 = q + p$

but $p + q = 1$

Therefore the sum of the terms in the column = q^2

Column headed Aa

$$2pq^2 + 2p^2q$$

Divide all through by $2pq = q + p$

but $p + q = 1$

Therefore the sum of the terms in the columns $= 2pq$

Column headed AA

$$p^2q + p^3$$

Divide all through by $p^2 = q + p = 1$

Therefore the sum of the terms in the columns $= p^2$

6

Genetic Linkage

6.1　General

Genes are said to be linked when they, or more accurately the loci (positions) which they occupy, are situated on the same chromosome. The reason for saying 'locus' rather than 'gene' is that at any given locus the gene may be one or other of a series of allelomorphs, e.g. either the blood group gene controlling A or B, or O.

Superficially it might be thought that since all the genes on a chromosome are inherited together, linked characters would be readily appreciated by looking at a pedigree, but it is clear on a little thought that the matter is not as simple as this. First of all, it is impossible to detect linkage unless two characters are segregating in a family. For example, no information could be obtained regarding linkage of, say, the genes for eye colour and the ABO blood group locus (to give an entirely hypothetical example) from a family segregating for blue and brown eyes but in which all the individuals were group O. When the matter of different allelomorphs (such as the ABO blood groups) is in question it will also be realized that a certain character may be linked with one of these allelomorphs, O for instance, in one family and with A in another. Crossing-over must also always be thought of, because occasionally genes will change places and two characters which are linked in several members of a family may become separated in other members because crossing-over has taken place in the formation of the parental gametes.

Much the best way to understand linkage is to study actual pedigrees and two of the early proved autosomal linkages will now be discussed in some detail. Sex-linkage is a different problem and has already been referred to in Chapter 2.

6.2　The nail-patella syndrome and the ABO blood group locus
(linkage discovered by Renwick and Lawler in 1955)

A syndrome is a group of abnormalities which constitutes a recognizable disease, and in the nail-patella syndrome there are malformations of the skeletal system consisting of absent or hypoplastic patellae (knee caps), dystrophy of some or all of the finger nails, abnormalities of the elbow joints and frequently the presence of iliac horns (small projections from the flat bones of the pelvis), these being usually demonstrable only on X-ray. Figure 6.1 shows some of these features. The condition is an ideal one for studying linkage for the following reasons: (*a*) It is inherited as an autosomal dominant and has never been known to skip a generation − it is therefore always detectable when present: (*b*) it carries little disability and does not shorten life and there are therefore often large families and several generations to study: (*c*) the striking nature of the condition means that family reports are usually accurate. It will quickly be realized that the other character studied, the ABO blood group, is also ideal, as everybody is either O, A, B or AB. However, to detect linkage the right type of family must be observed, this being the offspring of a double back-cross.

Example of a double back-cross mating

A double back-cross is a mating between one person who is heterozygous for the two characters under discussion and another who is homozygous for them. For example, such a mating would be that of a woman who was blood group A, but carrying O, and heterozygous for the nail-patella syndrome (the gene is so rare that she would not be a homozygote − this has never been reported) and a man who was skeletally normal and homozygous for group O (i.e. OO). Such a

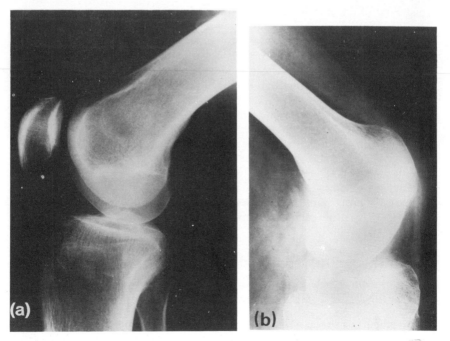

Fig. 6.1　The nail-patella syndrome. **(a)** Normal patella (knee cap). **(b)** Absent patella. **(c)** Dystrophy (abnormal growth) of finger nails. **(d)** Iliac horns on pelvis.

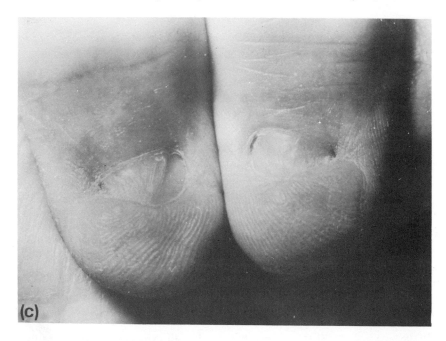

(c)

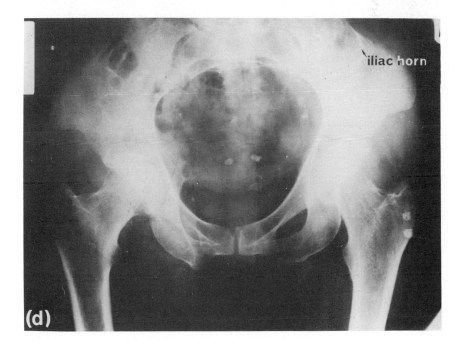

iliac horn

(d)

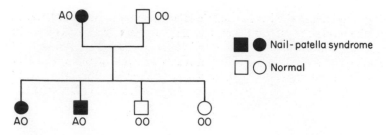

Fig. 6.2   Example of double back-cross mating.

situation is shown in Fig. 6.2, where the nail-patella syndrome has been inherited with blood group A.

It must be noted (*i*) that it is not possible to type the genotype AO, but it can be inferred from family studies, e.g. any person who types as A and has an O parent must be AO; (*ii*) that a much bigger pedigree than that shown would be necessary to *prove* the linkage.

Linkage masked by dominance

In the next pedigree (Fig. 6.3) part of a family is shown in order to demonstrate one of the elementary complexities of linkage studies. At first sight there might not appear to be linkage at all as in generation I the nail-patella syndrome appears in an A individual (I.1), in the next in two O people (II.1, the proposita, indicated by an arrow, and II.3), in the next in an A and an O (III.2 and 4), and in the next in an O (IV.1). In fact, however, the linkage is only masked by the dominance of group A. The proposita (II.1) has received the nail-patella gene

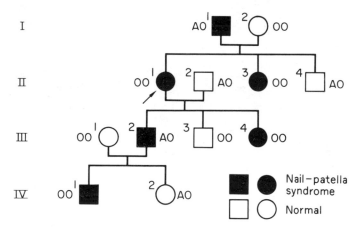

Fig. 6.3   Example of linkage masked by dominance. (Redrawn from Clarke, 1964.)

with her group O, and she has handed them on together to her son, III.2. As, however, the mother of III.2 was group O he must have received O from her and A from his father. He types as an A individual because A is dominant to O and the linkage of the syndrome with O is masked by this dominance of A. In generation IV it will be seen that III.2 has handed on the condition with O again. The situation is not due to crossing-over — if it were, IV.1 would have received the nail-patella deformity with A.

Crossing-over in linkage studies

Another pedigree (Fig. 6.4), one of our Liverpool families, shows what happens when crossing-over does take place. Here the nail-patella syndrome is linked with group A_1, as in I.1, II.1, II.5 and III.2.

Crossing-over has occurred at gamete (ovum) formation in I.1, after birth of II.5 and before the birth of II.7, who is, as will be seen, blood group A_1 and yet is not affected. It will be noticed that (apart from the propositus and II.7 already mentioned) the other A individuals in this family who are affected are A_2 and not A_1. (A_2 and A_1 are alleles and are distinguishable.)

The calculation of the cross-over value

The cross-over value, about 12%, in Fig. 6.4 agrees with that found in many families by Renwick and Lawler. It is calculated by looking at those individuals where the linkage can be assessed, i.e. II.1, II.3, II.5, II.7, III.1, III.2, III.5 and III.6 (III.3 and III.4 are not informative as neither parent was affected). In 1 case out of the 8 (12%) where the linkage could be detected a *new* combination is seen instead of the previous one, and this is quite close linkage. If the cross-over value reaches 50% it will be clear that there is an equal likelihood of the

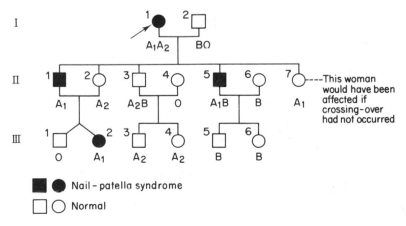

Fig. 6.4 Crossing-over in a nail-patella family.

characters occurring separately as there is of them occurring together, and therefore there is no linkage at all, but 'free recombination'.

Comparison of cross-over values in women and in men

It is of great interest that Renwick and Schultze (1965) analyzed the linkage data from 27 pedigrees of the nail-patella syndrome and found that crossing-over in respect of the nail-patella and ABO loci takes place twice as often in women (as it did in II.7, Fig. 6.4) as in men, and that in men it takes place less frequently as they grow older. This is the first investigation of its kind in Man, and is in accord with what happens at certain, but not all, loci in the mouse (in *Drosophila* it is well known that no crossing-over occurs in males). The implication of Renwick's sex difference findings might have to be considered in the mapping of chromosomes.

6.3 Elliptocytosis (ovalocytosis) and the Rh blood group system

Elliptocytosis is an abnormality of the red blood cells inherited as an autosomal dominant, the majority of the cells appearing elliptical, like Rugby footballs, instead of spherical (see Fig. 6.5). The condition is usually harmless though one form of it can give rise to anaemia, even in the heterozygote. Lawler and Sandler (1954) found that the gene responsible for elliptocytosis was linked to those controlling the Rhesus (Rh) blood group system, and it was later dis-

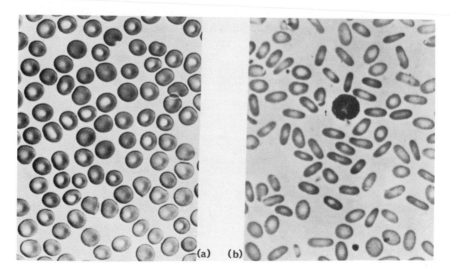

Fig. 6.5 **(a)** Normal red blood cells ($\times$ 500). **(b)** Elliptocytotic red blood cells ($\times$ 500). (From Carter (1962) by permission.)

covered that this linked gene was the harmless one, the other form of ellipto-cytosis (causing anaemia) *not* being linked to Rh. This knowledge is very useful clinically, since members of a family carrying the linked form can be told that their abnormality will cause no harm. Conversely, if someone is carrying the unlinked form they should be advised not to marry their first cousin in which case they might produce offspring more severely affected than themselves (the gene is so rare that they are most unlikely to marry a heterozygote in the general population).

The Rh blood group system seems extremely complicated but all that need be grasped here is that it consists of three pairs of allelomorphs – CDE and cde. These can occur in various combinations but the three loci are so close together on the chromosome (which is number 1) that they are always inherited as a unit.

In Fig. 6.6 it will be seen that elliptocytosis is linked to the genotype CDe: The propositus, II.1 is CDe/cDE, and she marries a man who is CDe/cde and therefore there is an 'affected' CDe chromosome and *also* an 'unaffected' one. Now it will be clear that any of the offspring who inherit cde must have obtained it from their father, and that those who have inherited cDE must have done so from their mother. In the third generation it will be seen that we can be sure of the origin of the Rh genes in III.2, III.3, III.4 and III.5, and that of these only III.4 has elliptocytosis.

III.3 (CDe/cDE) is not elliptocytotic because she received CDe from her father and cDE from her mother. For the fourth generation we do not know the full Rh genotype of the husband of III.4 (he is not shown in the pedigree), but we do know that he was probably heterozygous CDe as IV.2 has this genotype and yet is not affected. IV.2 therefore probably received cde from her mother

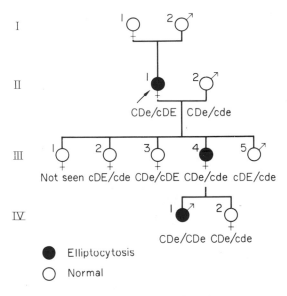

Fig. 6.6 Elliptocytosis and the Rh blood group system.

and CDe from her father. The elliptocytotic man IV.1 will have received his elliptocytotic gene from his mother with her CDe. The only alternative explanation for IV.2 being unaffected would be if there had been a cross-over in the gametes of III.4 between the births of her two children, whereby the gene for elliptocytosis got off her CDe and on to her cde, in which case IV.2 received CDe from her mother and cde from her father.

The first edition of this book stated 'It may come as a surprise to learn that the firmly established autosomal linkages in man are very few', but in the second edition (1977) pedigree analysis had gained impetus; now (1985) very many examples of linkage have been detected (see McKusick, 1985). Though pedigree analysis still remains the sheet anchor, yet cell fusion, and study of deletions, inference from the amino acid sequence of proteins and genetic engineering techniques have markedly increased the information available (see also chapter 12).

6.4   Detection of linkage by cell fusion

It has been known for a long time that certain virus infections of animals and man produce lesions in which cells with two or more nuclei are frequently found, and it is now known that this is due to the ability of the virus to fuse single cells together. The virus can be used experimentally to fuse different cell types from one species of animal, or cells from different ones. The hybrid cells combine the properties of the two parents, and have, to begin with, a double set of chromosomes. Later, however, there is selective loss, and the changes can be studied (the banding techniques are very helpful here see p. 57) along with concomitant alterations of enzyme synthesis. For example, mouse cells which cannot synthesize thymidine kinase were fused with human cells which could. After some generations all the human chromosomes but one (chromosome 17) were lost from the hybrid and yet it retained the ability to synthesize thymidine kinase. This indicated that the gene controlling thymidine kinase is located on chromosome 17 in Man.

Another use of cell fusion would be to replace a missing enzyme, a most important step forward – but only at present, potentially – in the correction of disease in Man.

6.5   Detection of linkage by the study of deletion

Certain disorders in man are due to chromosomal deletions. An example is the cri du chat syndrome, characterized by mental deficiency and a curious mewing cry due to weakness and under-development of the upper part of the larynx which is caused by a deletion of the short arm of chromosome 5. Suppose (entirely hypothetically) that a child with this syndrome had in addition a recessive condition such as blue eyes, inherited for example from an English mother who had married an Indian. One assumes that the Indian father was homozygous for brown eyes and therefore could not have been carrying blue. Brown eyes being dominant, the child should not have had blue eyes, and the fact that it had would mean that the lost piece of chromosome 5 also carried the locus controlling eye colour – in this case the 'brown' gene.

6.6   Detection of linkage by DNA probes

This is described in Chapter 12.

7

Association

7.1   Duodenal ulcer — clinical features

The duodenum (so called because its length is about the width of 12 fingers) is the first part of the small intestine; it adjoins the stomach from which it is separated by the pyloric sphincter and is continuous with the next part of the small gut, the jejunum. Ulceration, which simply means loss of continuity of the mucous membrane, usually occurs in the first part of the duodenum, for it is here that the acid-pepsin mixture from the stomach impinges, whereas in the second part of the duodenum the contents become alkaline. Duodenal ulcer (DU) is but one example of 'peptic' ulcer, this being liable to occur at any site where acid comes into contact with the mucous membrane of the gut, e.g. in the stomach itself, at the lower end of the oesophagus when the sphincter is lax, and in a congenital sac (Meckel's diverticulum) which sometimes persists in the small bowel and contains acid-secreting cells. Figure 7.1 shows X-ray pictures of a normal and an ulcerated duodenum after a barium meal; in the right hand picture the ulcer has penetrated into the deeper layers of the duodenum which therefore shows much scarring and deformity. The principal symptom of DU is pain two hours after meals, usually relieved by more food; occasionally there is bleeding into the gut, and sometimes perforation into the peritoneal cavity which leads to agonizing pain and board-like rigidity of the muscles of the abdominal wall.

7.2   Factors predisposing to duodenal ulcer — association with blood group O

Once a DU has developed it is quite certain that factors such as increased gastric acidity, smoking and anxiety may make the condition worse, but no one knows why an ulcer arises in the first place, since it is normal for the acid contents of the stomach to come in contact with the first part of the duodenum, many people smoke, and everyone from time to time has anxiety. Inheritance probably plays a part, patients with duodenal ulcer tending to have children who develop the same condition, but no simple method of inheritance is responsible. Support for a genetic predisposition is that there is a striking association with blood group O, particularly in those cases where the ulcer has

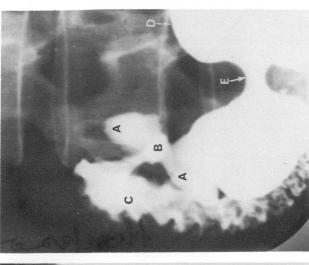

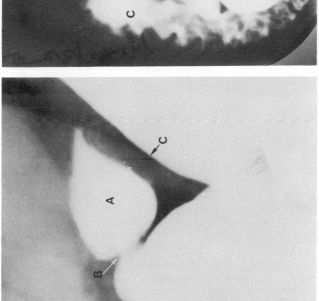

Fig. 7.1 Barium meal X-rays showing: (*left*) **A.** Normal duodenal cap (the first part of the duodenum). **B.** Pyloric canal. **C.** Lesser curvature of stomach. (*right*) **A.** Grossly deformed (because of scarring) duodenal cap. **B.** Barium filling an ulcer crater in the duodenum. **C.** Beginning of jejenum. **D.** Lesser curvature of stomach. **E.** Peristaltic wave. (By courtesy of Dr G. Scarrow.)

bled, and the main purpose of this section is to show how the validity of such an association may be tested.

7.3   Testing for an association between two characters

The association between blood group O and duodenal ulcer was established by collecting large numbers of patients with the disease and of normal controls, by the investigation of the ABO blood groups in both these groups and by the finding that there was a statistically significant excess of blood group O in the ulcer patients. Normally the controls consisted of healthy blood donors, students and nurses, but sometimes patients with diseases other than duodenal ulcer were used, though these are less satisfactory as there is always a possibility that the other diseases may be exerting some effect which has not been allowed for. The ulcer and control groups were then compared by means of the χ^2 test (see Bailey (1959) for the method, or any book on statistics) and the value of χ^2 was highly significant. For Liverpool the χ^2 found when patients and controls of O and not-O were compared was 44.26 with a probability of $< 10^{-10}$, i.e. the findings were enormously unlikely to have occurred by chance. An investigation of this sort often entails the consideration of whether it is legitimate to pool data from various localities, and it is not permissible to do this if the samples are heterogeneous, i.e. if the proportions of the characters being considered are significantly different from one another in the various samples which we want to pool. The χ^2 test for heterogeneity is worked out in exactly the same way as that for an association and if the value of χ^2 obtained is significant, it means that the figures in the various samples are heterogeneous and must not be pooled, but if it is not significant they can be. In the DU work all the data proved to be homogeneous and therefore poolable and the association was thus shown to be very highly significant.

7.4   Pitfalls in the selection of controls

While these results look conclusive, it is imperative to bear in mind a particular pitfall in the selection of controls, namely *racial stratification*. By this we mean that there is incomplete mixing of populations from different sources. An example is that among African Negroes a particular Rhesus blood group complex, known as cDe (see p. 24), is very common, though it is rare among Caucasians. In a mixed population it would be highly unlikely that there would be random mating between Negroes and Caucasians and therefore there would appear to be an association between dark skin colour and cDe. This is an extreme example, but less well-defined degrees of stratification are common. For instance, in this country there are Jewish communities who tend to marry among themselves, and clearly if an association were found between dark complexion and diabetes (a disease which is very common in Jews) this would be of racial and not genetical origin.

7.5   Method of testing for an association using sibs of patients as controls

While *a priori* racial stratification seems unlikely to explain the association between group O and duodenal ulcer, it *might* be operating, and to eliminate any possibility that it was doing so our team in Liverpool used, at the suggestion of Professor L. S. Penrose, the unaffected brothers and sisters (sibs) of the patients as the controls; and the details of the method are given below (see Clarke *et al.*, 1956).

(*i*) The sibships must segregate (i.e. separate) both for blood group and for ulcer/not ulcer.

(*ii*) The chance of the propositus being group O is calculated in each sibship separately. Suppose there are four sibs in one family, two of group O and two of not-O (e.g. A). One of these has an ulcer. Now the chance here of the ulcer patient being group O is clearly 50%, i.e. 0.5, and so 0.5 is the 'expected' value in this sibship. We now look up the records, and if we find that he is in fact group O we enter the figure for that sibship as 1. If, on the other hand, he is group A we enter the figure for that sibship as 0 (zero). It will be seen that this method takes into account the fact that sibships are composed of different proportions of groups O and not-O; for example, in a family consisting of 6 group O and one group A sibs the chance of the propositus being group O would be much greater than it would be in one consisting of 6 group A and one group O sibs.

When all the families have been assessed we have therefore a total 'expected' and a total 'observed' score and we have to work out whether there is a statistically significant difference between them. This is done as follows and entails finding the variance in each family:

(*a*) Divide the number of group O individuals in the family by the total number of individuals in the family. This will give the 'expected' number of ulcer cases of group O.

(*b*) Multiply this value by the number of sibs who are not group O divided by the total number of individuals in the family. This will give the variance of the 'expected' value.

In our example of the sibship of four in which two sibs are group O and two are group A it will be remembered that the chance of the propositus being O is $\frac{1}{2}$, and we can calculate that the variance is $\frac{1}{4}$ (see Bailey, 1959).

(*c*) Now take the sum of the 'observed' values and the sum of the 'expected' values, and find the difference.

(*d*) Next take the sum of all the variances and find the square root of it. This will be the standard error of the difference between the 'observed' and 'expected' – the 'observed' being, of course, the patients who actually are group O and the 'expected' being those who would be expected to be group O.

(*e*) Finally divide the difference between the 'observed' and the 'expected' by the standard error of the difference, that is, divide the value found in (*c*) by the value found in (*d*). *To be statistically significant the difference between the 'observed' and the 'expected' must be more than twice the standard error.*

To our surprise, after collecting about 160 segregating sibships, although we did find an excess of group O propositi observed over those expected, the difference was not statistically significant, but when combined with poolable data from the USA it became so, and the Establishment view now is that the association is a valid one. However, duodenal ulcer has become rarer and the association with group O has proved to be of no clinical importance. It has, therefore, been largely forgotten.

A final point that should be made is that in some diseases sibs are highly undesirable as controls (they give inaccurate information) see Huntington's chorea, p. 2, for an explanation of this point.

7.6   Association found to be due to a transfusion effect

Another pitfall which we discovered in Liverpool concerned an alleged association between duodenal ulcer and the Rhesus blood groups, but the association was found to be due to a *transfusion effect*, and this we think is so interesting that it will be described in detail. If individuals are simply scored as Rh positive and Rh negative no association is found between the Rh type and duodenal ulcer. However, it must be remembered that there are several different ways of being Rh positive because the Rh blood group is determined not by one gene but by *three* closely linked genes; each individual inherits two sets of these (derived from the two parents) and each set is inherited as a unit. For our purposes each unit consists of a combination of the well-known C, D, E, c, d and e antigens, all of which can be identified by the appropriate antisera, except for d, the antiserum for which has not yet been found (it is assumed that any individual who does not possess D is homozygous dd). It will be realized that as everyone inherits two units, they receive a combination of two from each of the C/c, D/d and E/e antigens.

Certain Rh positive combinations are commoner than others and we are accustomed to seeing particularly the following:

CDe/cde,   34.9%
cDE/cde,   14.1%
CDe/cDE,   13.4%

As regards the Rh negative combinations only those people who have two sets of cde are strictly Rh negative (rr) but in hospital practice, where the blood would normally only be tested with anti-D, individuals with a genotype such as cdE/cde, or Cde/cde would also be classed as Rh negative.

Buckwalter and Tweed (1962) *genotyping* their patients (i.e. not just testing for the presence or absence of the D antigen) found a highly significant association between the Rh positive combination CDe/cDE $(R_1 R_2)$ and duodenal ulcer and also blood type MN and this disease (in the MN blood group system the heterozygote MN can be detected serologically).

It occurred to our senior laboratory technician Mr W. T. A. Donohoe that *transfusion* might be the cause of the association since he had observed (in common with many other people who actually *do* the work) that, using the five routine Rh antisera, after a transfusion, mixed cell agglutinations were sometimes produced, indicating that the transfused cells contained antigens which the recipient lacked, these showing up as islets of agglutinated cells. An individual will be scored as positive even if only a proportion of the cells is agglutinated by the appropriate antibody.

From a consideration of the frequencies of the various Rh genotypes in the population it was clear that the combination CDe/cDE was just the one which would be increased if transfusion were the cause of the association with duodenal ulcer. Similarly with the MN blood groups the frequency of MN would be increased by transfusion.

Examination of Table 5.2 should make the matter clear, and the reader should pay particular attention to the last column where it will be seen what the patient gained in the way of antigens following the transfusion. It is of interest that such gains persist often for a month following transfusion.

We clinched the matter by testing duodenal ulcer patients who were in hospital (in either medical or surgical wards) and here we found the R_1R_2

Table 7.1 Blood groups before and after transfusion of the amount of blood stated. (From Clarke *et al.* (1962), by courtesy of the editor of *The British Medical Journal*.)

Case No.	Patient Pre-transfusion	Donor blood	Patient Post-transfusion
1	CDe/CDe MM	cDE/cde NN cDe/cde MM 2 pints (1140 ml) of blood given	CDe/cDE MM Has gained c, E, and N antigens
2	CDe/cde NN	cDE/cde MM 1 pint (570 ml) of blood given	CDe/cDE MM Has gained E and M antigens
3	cDE/cde MM	CDe/CDe MM CDe/cde MM 2 pints (1140 ml) of blood given	CDe/cDE MM Has gained C antigen
4	cDE/cde MM	CDe/cDE NN cDE/cde MN cDE/cde MN CDe/cde MM 4 pints (2275 ml) of blood given	CDe/cDE MN has gained C and N antigens
5	CDe/cde MN	CDe/CDe MM cDE/cde MN cDE/cde MN 3 pints (1700 ml) of blood given	CDe/cDE MN Has gained E antigens
6	cDE/cde NN	CDe/cdc MM CDe/cDE MN 2 pints (1140 ml) of blood given	CDe/cDE MN Has gained C and M antigens

association described by Buckwalter and Tweed — but over half of the patients had been transfused. By contrast in a series collected as out-patients who had had no recent transfusion the distribution of CDe/cDE did not differ significantly from that of the general population. A similar situation was found with respect to the MN groups (Clarke *et al.*, 1962).

Perhaps, when faced with an obscure association, as with an obscure diagnosis, one should increasingly think of the possibility — has treatment anything to do with it?

7.7   Cancer of the oesophagus and tylosis

In 1958 we discovered in Liverpool two remarkable families (and since then more have been described) in approximately half of whose numbers there was a skin condition (tylosis) which is characterized by thickening of the skin of the palms and soles. Tylosis is inherited as an autosomal dominant character. The point about the tylosis in these families is that those with the skin condition are extremely likely to develop a cancer of the lower end of the oesophagus (gullet) and in no instance has any member who is non-tylotic developed the cancer. Figure 7.2 shows part of one of the Liverpool families.

Common sense suggests that in these particular families the gene for tylosis also causes the stratified epithelium of the oesophagus to undergo a malignant change — in other words, the gene here has a pleiotropic effect which results in an 'association'. Alternatively, it is possible that we are dealing with two linked genes, one for tylosis and the other for the carcinoma, and that they are so

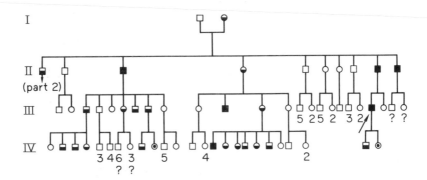

Fig. 7.2   Pedigree of cancer of the oesophagus and tylosis. (Redrawn from Clarke, 1964.) Numbers indicate number of individuals in any category throughout.

closely linked that no crossing-over has occurred. On our present knowledge this cannot be proved or disproved but if other similar families were found where the genes were in repulsion some members would have the cancer and some the tylosis. That is, the association would have disappeared and the evidence would strongly favour the existence of linkage.

7.8 Linkage disequilibrium

When the first edition was written we thought of linkage between two loci as being a chance occurrence and not implying causality, whereas association could be causal. However, 'linkage disequilibrium' gradually entered the genetic vocabulary and it expresses the idea outlined under 'the formation of a super-gene' (p. 24). In linkage disequilibrium linked alleles at different loci appear together more frequently than they would if they were segregating at random. This occurs particularly in the HLA system and may hint at causality in relation to the association with disease (see §7.9). However, there is a conundrum. Many of the diseases associated with HLA types have either a male *or* a female predominance and yet the linkage disequilibrium is the same in both sexes.

7.9 Association between HLA types and disease

The HLA system is described under 'transplantation antigens' on p. 28. Not unnaturally, with such a complex system many associations have been described. The strongest is that between an allele called BW27 at the B locus (W stands for 'Workshop') and ankylosing spondylitis. This is a disease principally of young men characterized by pain and rigidity of the back. Weaker associations have been found with disseminated sclerosis and with coeliac disease.

The biological mechanisms involved both in this and other HLA related diseases are not known but there are useful suggestions. Thus a given tissue antigen may be closely mimicked by a certain virus so that carriers of the allele fail to recognize the virus as 'non-self' and so have a poor immunological defence against it. Other alleles may confer the tendency to over-respond and thus produce excessive quantities of antibodies which cross-react with host tissues resulting in auto-immune disease. Whatever the explanation, there is no doubt that diseases and the HLA system are of great interest at the present time, and the famous question 'Why did he get the disease he did?' may be nearer solution.

There are a few *caveats* about HLA, particularly to be remembered as the journals are full of associations – and yet the techniques are difficult.

(*a*) Lymphocytes need accurately separating from platelets, since the latter contain HLA-A, -B, and -C antigens, and therefore may weaken the reaction of the antibodies.

(*b*) If test cells are homozygous for a particular enzyme antigen, they will have double the amount of that antigen as compared with heterozygotes. Because of this, homozygous B7 cells, for example, will react strongly to anti-B7 but (because of cross-reactivity) they will also react to anti-B27. It is thus difficult to decide whether the cells have both B7 and B27 or a double dose of B7.

(c) It is also essential to test lymphocytes with serum from the *same* population as that from which the donor was drawn. This is because different racial groups have widely different antigen frequencies and specificities, and consequently a serum may appear monospecific when tested on one population and yet be shown to possess extra specificities when tested on another.

When monoclonal antibodies (see Chapter 12) have been perfected, these will greatly help to solve the above problems.

7.10   A final thought on association

Teachers keep on repeating that association does not imply causality. True enough, but the tylosis story gives a nudge in that direction, though other associations have much less commonsense about them. The reciprocal is a useful point – if there is *no* association, causality can be definitely ruled out.

8
Chromosomes

8.1 General

Human somatic cells, no matter from which organ they are examined, contain 22 pairs of autosomes and two sex chromosomes, X and Y in the male and X and X in the female. It is now possible to examine and count chromosomes but they can only be studied in actively dividing cells and therefore the bone marrow is often looked at or cultures made from the lymphocytes in the peripheral blood or from skin. Whichever type of tissue is used the dividing chromosomes appear as is shown in Fig. 8.1 and it must be appreciated that

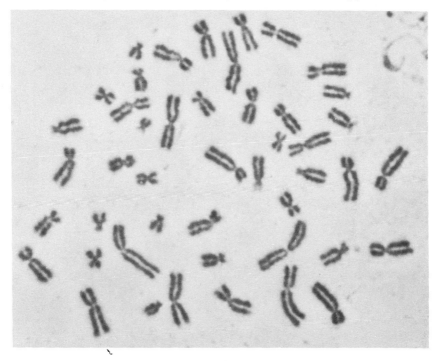

Fig. 8.1 The 46 chromosomes from a single male somatic cell undergoing mitosis. The chromosomes have doubled but are still held together by their centromeres. (By courtesy of Dr S. Walker, Cytogenetics Unit, Royal Liverpool Hospital.)

what we are looking at are chromosomes in mitosis not meiosis. The dividing chromosomes are known as chromatids; they are held together at a sort of 'waist' (the centromere) and this is located near the centre of some chromosomes, which are then referred to as metacentric, and near the end of others (acrocentric). Many chromosomes therefore have a long and a short arm.

Figure 8.2 shows how by convention the 22 pairs of autosomes are arranged and numbered in decreasing order of size. The X and Y are not numbered.

Meiosis (reduction division) takes place during the formation of the germ cells and it is then that the chromosome number is halved so that each sperm and each egg only contains 23 chromosomes. During meiosis, crossing-over of genetic material occurs between homologous chromosomes so that no sperm or ovum is exactly like any other. Crossing-over provides the basis for variability in a species. *Note* It does *not* take place when the chromosomes of the egg and spermatozoon come together at fertilization.

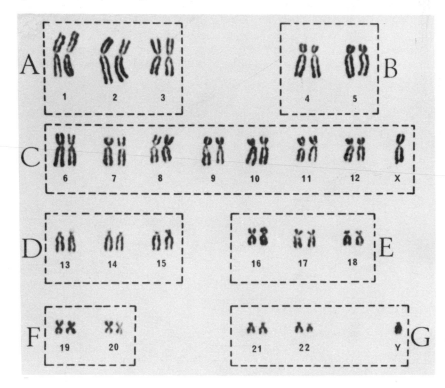

Fig. 8.2 The same 46 doubled chromosomes as in Fig. 8.1. The chromosomes are arranged in decreasing order of size and numbered from 1 to 22. The X and Y are not numbered. The letters A to G show the various groupings; individual chromosomes in any of the groups can now be recognized by banding techniques. (courtesy of Dr S. Walker, Cytogenetics Unit, Royal Liverpool Hospital.)

8.2 Staining techniques

By means of special staining techniques (either fluorescent staining viewed with an ultra-violet illuminated microscope or by Giemsa after pre-treatment) all the pairs of human chromosomes can now be individually recognized by their banding patterns, and this is of great value in identifying abnormalities and in linkage studies where previously there may have been difficulty in identifying a particular chromosome. (Giemsa is a well known stain and would normally stain the chromosomes uniformly blue, but with pre-treatment by

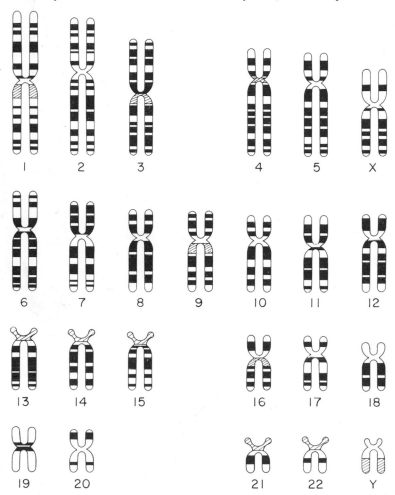

Fig. 8.3 Diagrammatic representation of banding patterns of individual chromosomes as revealed by fluorescent and Giesma staining (see *Paris conference*, 1971, supplement 1975 and also *Cytogenetic Cell genetics*, 1978, **21**, 313,404. (From Emery, 1983, by courtesy of the author.)

trypsin each chromosome shows a characteristic banding pattern.) Figure 8.3 shows a diagrammatic representation of the banding patterns the (male) chromosomes being arranged in pairs, and in groups.

8.3   The Barr body and Lyonization of the X chromosome
(Lyon, 1961)

Examination of the chromosomes takes time and where abnormalities of the sex chromosomes are suspected useful preliminary information can often be obtained by looking at a stained preparation of cells from the inside of the patient's mouth. In the normal female can be seen a small darkly staining body under the nuclear membrane which is not present in males. It is known as the Barr body (see Fig. 8.4) and its presence means that there are two X chromosomes in the nucleus, only one X chromosome (either the maternal or the paternal one) being active in any one cell and it is the inactive one which stains darkly (Lyonization). In males, because there is only one X chromosome in each cell, this is always active and there is no Barr body. People who have one or

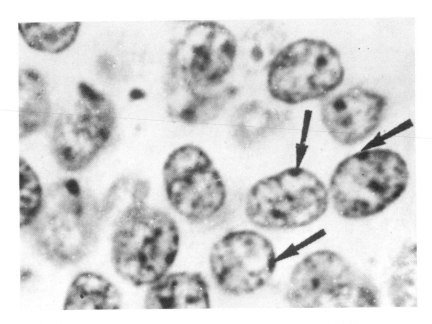

Fig. 8.4   Epithelial cells from the buccal mucosa of a female patient showing Barr bodies in the nuclei (chromatin positive). (Redrawn from Clarke, 1964, by courtesy of Dr Winston Evans.)

more Barr bodies are known as 'chromatin positive' and those who have none as 'chromatin negative'. It is not normal to have more than one Barr body, or to have none at all, if you are a woman. If you are a man, it is not normal to have one.

One difficulty, however, with Lyonization concerns a blood group system which is X-linked (the Xg system). Individuals are either Xg^a positive or negative, identified by anti-Xg^a antibody. Women with normal X chromosomes who are heterozygous Xg^a/Xg, having received Xg^a from one parent and Xg from the other, ought according to Lyonization to have two races of red blood cells, one possessing the Xg^a antigen and the other not. This, however, is not so – they have the antigen on *all* their red cells. It seems that the X is not inactivated over the whole of its length when it is normal, though it is when it is abnormal (e.g. an isochromosome). By contrast in G6PD deficiency (p. 71) there *are* two races of red cells in a heterozygous female. This is because the gene is on a different part of the X.

Lyonization also explains why 15% of carrier females of the haemophilia gene have a normal concentration of factor VIII (see p. 8).

No one knows exactly at what stage of development Lyonization occurs, but suppose that it happens at the 8 cell stage of the embryo. At that time, which of the two Xs is inactive in a given cell is a matter of chance, similar to tossing eight coins at one time in the air. If one performs this trial many times, representing different individuals, the most common result in the long run would be four heads and four tails. However, all the other different combinations will occur, though less frequently. Exceptionally, there would be eight heads (or eight tails). Translating this model back to haemophilia, it can be appreciated that in an exceptional female carrier a normal quantity of factor VIII is produced – but she is still a carrier, and one of the 15% who would be missed.

8.4 International nomenclature

Because so much more is known about chromosomes there is now an agreed series of abbreviations which can describe all types. Normal chromosomes are written as 46,XX if female and 46,XY if male. The letter p stands for the short arm of a chromosome and q for the long arm; t stands for 'translocation' and i for 'isochromosome'. A plus sign *before* a number or group means that there is an extra one of those particular chromosomes and a minus sign that one is missing – in both cases a whole chromosome. When the plus or minus sign is *after* the number (or group) it means that part of that chromosome has either been added to or deleted. The appropriate abbreviations are given for the abnormalities described below. These are simple ones and the nomenclature can be much more complicated in other cases. It will be remembered that the sex chromosomes are not numbered so the numbers refer to autosomes.

8.5   Some conditions associated with chromosome defects

The disorders to be described all have karyotypes (see glossary) which are abnormal in number, in one there being an abnormal number of autosomes and in three, an abnormal number of sex chromosomes. On occasion, however, two of the conditions may also be produced by structural defects.

Down's syndrome (mongolism), caused by the presence of an extra small autosome

This is named after Langdon-Down, who first described the condition in 1866 and though it is still sometimes known as mongolism owing to the slanting eyes and slightly flattened face of the patients, this is for obvious reasons of international courtesy not such a suitable name. (In any event, it is of interest to note that the Mongoloid races think the patients have a European appearance and the resemblance to the Oriental is at most superficial; in fact it is quite easy to

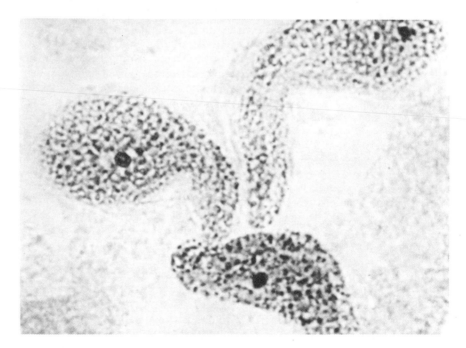

Fig. 8.5   Nuclei of tissue cells containing heteropyknotic body from a larva which developed into a female *H. bolina* butterfly (× 90 oil immersion objective).

recognize Down's syndrome in patients of Oriental racial origin.) As well as the curious facial configuration and the narrow eyes, the patients are very short and there is mental retardation which varies from mild to severe, but the children often have a happy and affectionate disposition. In newborn babies the condition can be recognized because the head tends to be small and oval, the ears low-set and with small lobes and the eyes slanting upward and outward; the bridge of the nose is usually absent or poorly developed and the mouth tends to hang open with the tongue out. Grey-white specks are seen in the iris of the eye (Brushfield's spots) and the little fingers are often short and incurved; the hands are broad and there are abnormal finger and palm print patterns. The life expectancy used to be about eight years, the children dying from infections, from heart defects and being more than usually liable to leukaemia. However, nowadays with the introduction of antibiotics and heart surgery they live much longer. Table 8.1 shows the frequencies of the syndrome at various maternal ages (and also the outlook for subsequent children) and it will be seen that the older mothers are much more at risk than the younger ones. The age of the father does not seem to be a factor.

The usual type of Down's syndrome is caused by non-disjunction (see glossary) of the small chromosome pair 21. The patient is trisomic for this and has a complement therefore of 47 instead of 46 chromosomes, and the technical name for the condition is trisomy 21 (47,XX, +21 if female, 47,XY, +21 if male).

Trisomies are produced by several types of error in cell division. The most common is known as meiotic non-disjunction, which means that a pair of chromosomes has failed to separate during the production of sperm or egg cells, more usually the latter. This results in one cell containing both members of a particular pair of chromosomes and one containing neither. If the one with the extra chromosome joins with a normal cell (ovum or sperm) at fertilization, the resulting child will have three of that chromosome instead of a pair, and will be said to be trisomic for that chromosome (he will have 47 in all). If on the other hand the cell which lacks a chromosome unites with a normal cell, the resulting individual will be monosomic for that chromosome (and will have 45 in all). Monosomy as regards the autosomes is usually, if not always, lethal.

Table 8.1 Relationship of maternal age to trisomy 21. (From Redding and Hirschorn, 1968, by courtesy of the authors and the editor of the March of Dimes Original Article Series, Vol. IV, no. 4.)

Maternal age	Risk of occurrence	Risk of recurrence
20–30	1 : 1500	1 : 500
30–35	1 : 750	1 : 250
35–40	1 : 600	1 : 200
40–45	1 : 300	1 : 100
45–up	1 : 60	1 : 20

Very occasionally 'Down' women have babies, and as might be expected, half their children are Down and half normal.

The second type of error, which occurs much less frequently, is caused by a translocation of Robertsonian type (see glossary) between a number 15 and a number 21 chromosome. Figure 8.6 demonstrates the method of inheritance of this type and the details are as follows.

There is fusion of part of a chromosome 21 and part of a chromosome 15, but an individual carrying this abnormality will not be affected, because, even though he has an abnormal chromosome in the sense that he has parts of two stuck together, he still has no excess of chromosomal material. Those of his children, however, who receive his abnormal chromosome and *also* receive his normal 21 will be effectively trisomic, because they will have received a normal chromosome 21 from their other parent.

This type of Down's syndrome occurs in children of younger mothers, and it will be appreciated that it is not dependent on a chromosomal accident due in part to maternal age, but is brought about by the direct inheritance of an abnormal chromosome. Wherever there is a family history of Down's syndrome in relatives, or where a young mother has already given birth to one and is likely to have more children, it is well worth examining her chromosomes and those of her husband, and it would be very much the duty of a doctor to have such an investigation carried out. If a translocation is found in one of the parents, it will be seen from Fig. 8.6 that there is an even chance of the child inheriting the translocation. If it inherits *both* the translocation *and* a chromosome 21 from its carrier parent, it will have Down's syndrome because it has extra chromatin, the other 21 being derived from the normal parent. Abnormal gametes are, however, formed less frequently than normal ones, especially in males, so that the outlook when the father is a translocation carrier is less bleak than would be thought from the above.

Mosaicism in Down's syndrome

Sometimes non-disjunction occurs after the zygote has been formed. It is then mitotic in origin and errors can occur at any division, so that two or more cell lines may be established, and it is this which is known as chromosomal mosaicism (see Fig. 8.7). Mosaicism in Down's syndrome is of particular interest to the doctor since not infrequently one sees individuals who, though perfectly normal mentally, have some of the other physical stigmata of the disorder, i.e. abnormal palm print patterns. The severity of the symptoms of Down's syndrome is directly related to the proportion of trisomic cells in the mosaic individual. The possibility that parents who have had more than one 'regular' (i.e. trisomic) 'Down' child may themselves be mosaics must be considered. Such parents might appear normal or near normal but they could produce some normal and some 21 trisomic gametes, thus greatly increasing the risks to their children.

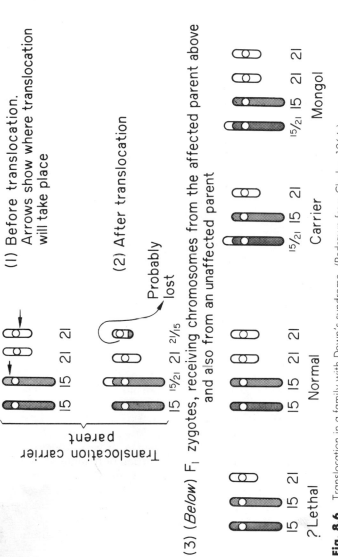

(1) Before translocation.
Arrows show where translocation
will take place

(2) After translocation

Probably
lost

Translocation carrier
parent

15   15/21   21   21/15

(3) (*Below*) F₁ zygotes, receiving chromosomes from the affected parent above
and also from an unaffected parent

15   15   21   21   15   15   21   21   15/21   15   21   15/21   15   21   21
        Normal   Carrier   Carrier   Mongol

15   15   21
  ?Lethal

Fig. 8.6   Translocation in a family with Down's syndrome. (Redrawn from Clarke, 1964.)

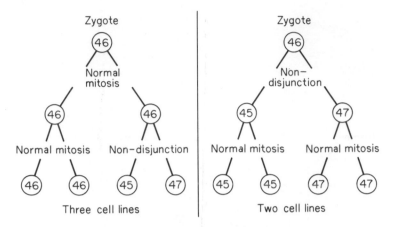

Fig. 8.7   The production of chromosome mosaics by mitotic non-disjunction after formation of a normal zygote. (From *Chromosomes in Medicine*, Ed. J.L. Hamerton, 1962. By courtesy of Dr D. G. Harnden and Heinemann Ltd.)

Other (rare) chromosomal abnormalities involving the autosomes

Trisomy of the 14–15 D and 16–18 E groups of chromosomes has also been described and this again produces mental defect. A deletion of part of chromosome number 5 is responsible for the cri du chat syndrome (46,XY, 5p-) (or XX as the case may be) referred to in another connection (see p. 44).

Klinefelter's syndrome(47,XXY), caused by the presence of an extra X chromosome in a male

These patients look and behave like males, and it is highly probable that a proportion of cases go through life without ever consulting a physician, but they are completely infertile as even though erections can be obtained and ejaculations occur, spermatozoa are never present and the fluid is derived from the prostate and accessory glands. Some patients are married and consummation can be legally effected. The patients have small testes, are often tall and thin and have high-pitched voices. They may have abnormal development of the breast tissue and it is often only this which brings them to the doctor. In addition, there is a liability to mental defect, though many patients have normal intelligence.

The defect is caused by the fact that patients have an extra X chromosome, due to non-disjunction having occurred in one of the parents. They have therefore 47 chromosomes in all. Figure 8.8 shows the results of non-disjunction of the sex chromosomes occurring (a) in a woman and (b) in a man.

These offspring, as can be seen, have an abnormal sexual constitution as follows:

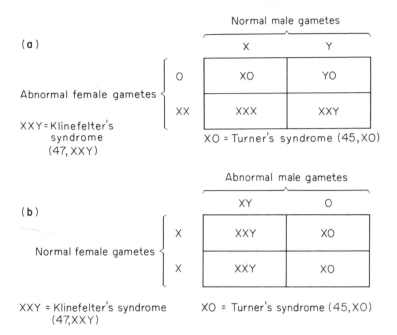

Fig. 8.8

XO   Abnormal female (Turner's syndrome)
YO   Almost certainly inviable (possibly because the X chromosome carries the gene controlling blood clotting factors)
XXX Abnormal female (triple X). A rare condition and the patient is often surprisingly normal
XXY Abnormal male (Klinefelter's syndrome)

The abnormal gametes in Fig. 8.8 are produced if the non-disjunction occurs at the first meiotic division. When it occurs at the second meiotic division (which is effectively mitotic) sperm containing 2 XXs or 2 YYs are produced (or conversely neither). Then at fertilization XYY males can result (see below), YY from the father and a normal X from the mother.

Patients with Klinefelter's syndrome are chromatin positive (see p. 58), that is they have a Barr body because they have two X chromosomes. It has been found by means of family studies, using the X-linked Xg blood group (see p. 12), that in Klinefelter's syndrome the non-disjunction has taken place in the father in about 40% of cases and in the mother in about 60%.

The XYY syndrome

Men with 47,XYY chromosomes are on the whole taller than average. Earlier reports suggested that the condition was of a high incidence in maximum security prisons, but it is now doubtful whether the association is very strong between this abnormality and dangerous or violent propensities. The prevalence of this chromosome abnormality at birth is about 1 in 700 newborn males, and as there is no evidence of an increased mortality in childhood it seems likely that most individuals who are 47,XYY lead normal lives. If therefore the condition is found incidentally as a result of amniocentesis (see p. 94) it becomes a problem as to what to tell the parents.

Turner's syndrome (XO), caused by the lack of one of the two X chromosomes (45,XO)

The characteristic picture of a patient with this condition is that of a very short girl having primary amenorrhoea (i.e. not having begun to menstruate) and who lacks the secondary sexual characteristics. There is often webbing of the neck (the skin filling in the angle of the neck and shoulders) and an increased carrying angle of the forearm, i.e. the forearm is 'set off' more than normal from the upper arm.

Most patients on skin sexing are chromatin negative (because the second X is lacking), and chromosome counts confirm that only 45 chromosomes are present, the missing one being an X and the condition arising because of non-disjunction (see Fig. 8.8). Family studies using the Xg X-linked blood group system have shown that the non-disjunction occurs more frequently in the father than in the mother in contradistinction to Klinefelter's syndrome (see above).

Turner's syndrome caused by mosaicism, or by the formation of an isochromosome (46,Xi,(Xq) or 46,Xi,(Xp))

As with mongolism, however, the situation is not entirely clear-cut, and Turner mosaics, e.g. XO/XX, can occur and this explains why some of the cases (one-fifth of the total) are chromatin positive, i.e. have a Barr body. Another reason is that the syndrome can also result from the single normal X being partnered by an abnormal X, that is an isochromosome, and Fig. 8.9 shows how this is formed. Sometimes the abnormal X is made up of two short arms and sometimes of two long ones, and although the individual has two XXs and therefore a Barr body, yet she lacks some of the genes which make her a normal female, i.e. those on the missing long or short arms. A consideration of many cases with both types of isochromosome suggests that genes present on the long and the short arm of *both* X chromosomes are necessary for the development of a functioning ovary, whereas genes apparently influential in the development of stature, and of the other somatic features which are frequently abnormal in Turner females are on the short arm, and that the short arms of *both* Xs are needed for normal somatic development.

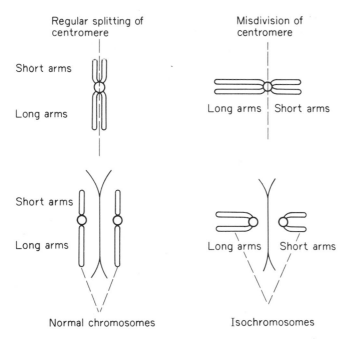

Fig. 8.9 The origin of an isochromosome through misdivision of the centromere during mitotic division. (From *Chromosomes in Medicine*, Ed. J. L. Hamerton, 1962. By courtesy of Dr D. G. Harnden and Heinemann Ltd.)

Chromosomes and cancer

Many tumours have abnormal chromosomes but whether they are the result or the cause of the malignancy is arguable. In favour of the first view is that chromosome abnormalities may be much more marked in a metastasis (see glossary) than in the primary growth. There is more evidence for the second view.

(*a*) In normal diploid mammalian cells life-span is limited whereas cells derived from malignant tissue appear to be immortal. Normal mammalian cells can be induced to escape from their commitment to senescence by exposing them to some mutagenic carcinogens, though the resulting capacity for indefinite multiplication precedes, and may be necessary for, malignant transformation (see Ph[1] chromosome below).

(*b*) Treatment by ionizing radiation for various diseases, e.g. ankylosing spondylitis (see p. 53) results in abnormal chromosomes and a proportion of the patients develop leukaemia, but it is uncertain whether the leukaemia cells arise from the irradiated ones.

(*c*) In a particular type of leukaemia (chronic myeloid) there is a very small chromosome 22, its long arm having been translocated to the long arm of chromosome 9, i.e. t(22q−; 9q+); 22q− is known as the Philadelphia (Ph[1]) chromosome after its place of discovery. In remission it disappears from the blood, but not from the marrow, it sometimes appears before the disease develops and patients are not born with it. Curiously the prognosis is better in Ph[1] positive than in Ph[1] negative chronic myeloid leukaemia.

An interesting finding relating to ionizing radiations in Man is that although the incidence of leukaemia was much increased in the survivors from Hiroshima and Nagasaki, the incidence of congenital malformations in the offspring of the irradiated population was no higher than normal, suggesting that the germ cells are more resistant than somatic cells.

(*d*) Cell fusion has also been applied to the study of cancer. Henry Harris at Oxford (1970, 1971) has fused highly malignant mouse tumour cells with non-malignant mouse cells, and then transplanted the hybrid cells into mice which had been irradiated to prevent them from rejecting the transplant (irradiation inactivates the immune response). Instead of a tumour forming in 100% of the mice, which is what happens when the malignant cells alone are injected, tumours developed in *only a third* of them – that is, cancer did not occur as long as the hybrid cells retained the full double set of chromosomes – one whole set from each parent. When, after dividing, they gradually started to lose chromosomes, the hybrid cells began to become malignant again. Selective chromosome loss may therefore be a factor in causing cells to become malignant.

(*e*) Certain rare autosomal recessive disorders have a high predisposition to malignancy, sometimes because they repair DNA inefficiently. In a condition known as xeroderma pigmentosum there is defective repair following damage by ultraviolet light, and the progressive accumulation of unrepaired DNA seems to induce mutations in somatic cells which may underlie the origin of cancer in the sun-exposed skin.

(*f*) It has been known for many years that viruses can produce cancer in animals and it was thought that the virus inserted some of its own DNA sequences which disrupted the activities of the host cells, so leading to cancer. Much more recently it was found that healthy cells possessed DNA sequences which were very similar, if not identical, to those of the cancer-producing viruses. This discovery gave rise to the idea that genes producing cancer – 'oncogenes' – had originated in the genetic material of normal cells and that somehow during their evolution viruses had incorporated them into their own genome – a complete *volte-face* of the original theory.

Obviously the oncogenes present in every cell in the body only cause cancer rarely – suppressor genes keep them in check. They exist in the first place as 'proto-oncogenes', and many stimuli, for example hydrocarbons such as methylcolanthrine, are capable of causing the change. They may do this by causing a mutation in the suppressor, or a chromosome translocation may bring about a change in the position of a proto-oncogene so that it comes under the influence of a different 'promoter' (controlling switch region of DNA). In Burkitt's lymphoma (a common cancer of lymph glands in Africans) the *myc* oncogene is translocated on its piece of chromosome to a position adjacent to a very potent promoter which normally controls the immunoglobulin genes responsible for making antibodies. About 25 oncogenes have been identified.

Finally we are back to where we started, for viruses themselves can activate oncogenes.

(*g*) There is an astonishing chromosomal abnormality in a tumour of the trophoblast known as a hydatidiform mole. What happens is that the abnormal ovum loses its own nucleus and is fertilized by an X sperm which then duplicates, hence all the inherited characters of the mole are paternal but a Barr body is present because there are two Xs. This finding was discovered by two Japanese, Kajii and Ohama in 1977.

9
Pharmocogenetics

Pharmacogenetics was originally defined as a study of genetically determined variations in animal species which are revealed solely by the effects of drugs. Loosely, however, it now includes hereditary disorders in which the symptoms can occur spontaneously but are often precipitated or aggravated by drugs. Examples are now given in which drugs, in one way or another, are involved in genetic problems.

9.1 Acatalasia — clinical features

This is an extremely interesting disease originally described in Japanese families, affected individuals being liable to severe ulcers in the mouth. To understand the condition it is necessary to remember the following points:

(a) Hydrogen peroxide dropped on to a raw surface in a normal person froths and the blood does not alter in colour. This is because the peroxide is degraded by an enzyme in the red cells and in the tissues called catalase, and oxidation of haemoglobin by hydrogen peroxide is thereby prevented.

(b) In patients with acatalasia frothing does not occur and the tissues turn brown because haemoglobin *is* oxidized.

(c) Certain bacteria (particularly some known as haemolytic streptococci) themselves produce hydrogen peroxide, and in a patient lacking catalase the haemoglobin of the blood reaching any tiny abrasion in the mouth is oxidized so that death of the tissues occurs because the infected area is deprived of oxygen. In this situation the bacteria multiply, the hydrogen peroxide production increases and a vicious circle is established.

Only about half of those who lack the enzyme actually show symptoms and those who do usually exhibit their disability before the age of 10 years, the bones of the jaw becoming infected, this leading to the loosening and loss of teeth. However, once all the teeth have been removed the ulcers heal and many patients remain permanently free of symptoms. A similar condition has also been found in certain breeds of dog and in guinea-pigs and mice.

Method of inheritance of acatalasia

There is good evidence that the trait behaves as an autosomal recessive character. Consanguinity is frequent and not a single parent of an affected individual has had the condition. Catalase estimations on patients and their sibs usually give a trimodal distribution, the intermediate values being those of the presumed carriers of the trait. However, this is not always the case since there are several forms of the condition.

More recently Swiss workers have found two examples of acatalasia on screening 18 459 blood samples from Army recruits and it is of interest that in neither case was there any clinical abnormality.

9.2 Primaquine sensitivity (G6PD deficiency)

Haemolytic anaemia, that is anaemia produced by the breakdown of red blood cells and not by haemorrhage, was recognized as an occasional complication of an antimalarial drug, pamaquine, when it was introduced into medicine in 1925. At first the anaemia was thought to be due to a hypersensitive or immune mechanism, but an antibody was not discovered and no explanation could be found for the occurrence of haemolytic anaemia in these sensitive subjects.

With the advent of widespread malaria treatment with the very similar drug primaquine during World War II more cases of a similar haemolytic anaemia were studied.

Clinical features of primaquine sensitivity

When a sensitive subject is given 30 mg primaquine daily he does not develop haemolytic anaemia for two or three days. Thereafter his urine gradually turns dark, muscular pains occur and anaemia and possibly jaundice appear. Discontinuing the drug results in a return to normal over a few weeks. However, if the symptoms are not severe and primaquine ingestion is continued he will, surprisingly, also gradually improve. It will be seen later that this is a most important observation.

Labelling experiments

Light was thrown on the mechanism of the anaemia when Cr^{51}-labelled red cells were transfused from sensitive subjects into non-sensitive recipients. The survival of these cells was normal until primaquine was administered, when they became lysed i.e. disintegrated. When Cr^{51}-labelled cells were transfused from a normal subject into a sensitive individual, they survived normally, even when primaquine was administered, and even when the patient's own red cells were lysed under the influence of the drug.

The fact that a sensitive subject gets better despite continued drug administration was investigated by similar methods. Thus, selected labelling of red cells of narrow age range with Fe^{59} resulted in the finding that red cells in a sensitive subject can be lysed by primaquine when 63 to 66 days old but not when 3 to 21 days old. Therefore, it seems that it is the ageing red blood cells which are destroyed by primaquine. Spontaneous clinical recovery while continuing with drug ingestion is due to the regeneration of a red cell population with a low mean age.

Biochemical studies

The normal red cell is known to possess enzyme systems which are concerned with the metabolism of glucose. One of these is glucose-6-phosphate dehydrogenase, *which is diminished in primaquine-sensitive individuals.* The first observation which led to this discovery concerns reduced glutathione. Thus, when both sensitive and non-sensitive red cells are incubated with primaquine

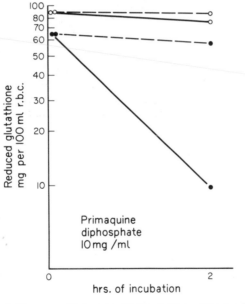

o— — —o Non–sensitive red cells incubated without primaquine

o————o Non–sensitive red cells incubated with primaquine

•— — —• Sensitive red cells incubated without primaquine

•————• Sensitive red cells incubated with primaquine

Fig. 9.1   The effect on reduced glutathione of incubating sensitive and non-sensitive red cells in the presence of glucose. (Beutler *et al.*, 1960, by kind permission of the McGraw-Hill Book Co. In *The Metabolic Basis of Inherited Disease*, Ed. Stanbury, Wyngaarden and Fredrickson.)

Site of metabolic defect in primaquine sensitivity

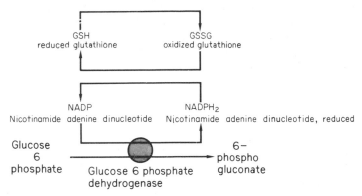

Fig. 9.2 The hydrogen atom removed from G6P by G6PD is taken up by NADP and $NADPH_2$ is formed. This in turn reduced GSSG to GSH. If G6PD is deficient, the cycle is interrupted and much less GSH is formed.

in vitro, their content of reduced glutathione falls (Fig. 9.1), but this can be prevented *in normal individuals only* by adding glucose to the buffer in which the red cells are suspended. The reason for the continued fall in sensitive red cells is due to a fault in glucose metabolism, ultimately the result of a defect in glucose-6-phosphate oxidation (dehydrogenation) due to a deficiency in the appropriate dehydrogenase (G6PD) and Fig. 9.2 explains the metabolic defect in relation to glutathione.

Genetic studies

Males are readily assignable to either the sensitive or non-sensitive group on both glutathione and G6PD studies. Women do not give quite so clear-cut a division and intermediate values are observed with either method of assessment because of the random inactivation of one X chromosome in each cell (see p. 58). They have in fact two populations of red cells.

On investigating 30 sibships it was found that male to male transmission was rare, appearing only once. The usual picture was that of an 'intermediate' female who produced sensitive sons or 'intermediate' daughters (cf. haemophilia, see p. 9).

Pedigree analysis and the sex differences led to the conclusion that the gene controlling the presence or absence of G6PD is on the X chromosome, i.e. the trait is a sex-linked 'dominant', the penetrance of the gene being incomplete because 'intermediate' daughters and sensitive sons are sometimes found to have apparently normal parents. Colour vision studies also established that G6PD deficiency was sex-linked and investigations with the Xg[a] antibody (see p. 12) then showed that the gene was close to the Xg locus which itself is probably on the short arm of the X chromosome.

Different forms of G6PD

Modern methods of protein separation such as electrophoresis have revealed scores of different forms of the enzyme G6PD. Some of these variant forms have normal enzymic activity while in others this is extremely low. It is the latter class – due to structural changes in the enzyme – which gives rise to the phenomenon of primaquine sensitivity. Some of these variant forms also give rise to spontaneous mild haemolytic anaemia.

G6PD and malaria

There is now substantial evidence from Nigeria that heterozygous females and G6PD deficient males possess resistance to the lethal form of malaria caused by *Plasmodium falciparum* (malignant tertian malaria). This is shown by their having lower parasitaemia rates than normal individuals and would seem to explain the finding that the geographic distributions of G6PD deficiency and malaria on a global basis are very similar. It is a curiosity of medical science that a phenomenon discovered initially by means of a side effect of an anti-malarial drug should have led to an understanding of a biological protective mechanism against endemic malaria.

Drugs to avoid

Primaquine is still a very important antimalarial drug but is obviously not commonly used in the UK or other temperate countries. However now that there are large immigrant communities with a significant incidence of G6PD deficiency, it is important to note that many other drugs can cause haemolysis in G6PD deficient individuals. Examples are:

Drug	*Used to treat*
Sulphapyridine (from salicylazosulphapyridine)	Inflammation of the colon *viz*. ulcerative colitis and Crohn's disease
Nitrofurantoin	Urinary tract infections
Chloramphenicol	Typhoid and severe meningitis
Nalidixic acid	Urinary tract infections
Quinidine	Cardiac arrhythmias

9.3   The metabolism of isoniazid

Family studies

Isoniazid is a drug, discovered about 1952, which is extensively used in the treatment of pulmonary tuberculosis. It had been known for a long time that people were either 'good' or 'poor' excretors according to whether they

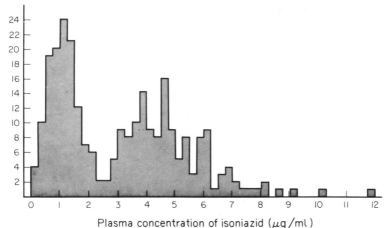

Plasma concentration of isoniazid (μg/ml)
6 hrs. following oral administration of 9·7 mg per kg of body weight

267 family members — 53 families

Fig. 9.3 Distribution of plasma isoniazid concentration. The rapid inactivators (R) are to the left of the antimode and the slows (S) to the right. (From Evans *et al.*, 1960, by courtesy of the authors and the editor of *British Medical Journal*.)

excreted large or small amounts of the compound in the urine and this led to the suggestion that there were, in fact, two genetically controlled classes in the population, the 'rapids' and the 'slows' and a pair of autosomal genes were thought to be responsible, rapid inactivation being dominant to slow. However, in the early work there was a considerable overlap in the two classes so that the scoring of some individuals was doubtful. Evans and his colleagues (1960) therefore carried out a much bigger investigation and used a more accurate method of assay. Taking 484 people they estimated the plasma concentration of isoniazid six hours after a single oral dose of 10 mg per kg body weight, and found a bimodal distribution with an antimode at 2.5 micrograms/ml. Of the 484 individuals studied, 267 were members of 53 complete, two-generation, Caucasian family units, and Fig. 9.3 (see above) shows the isoniazid distribution histogram for these subjects. It will be seen that approximately half were slow (S) and half rapid (R) inactivators, and analysis of the mating types confirmed that the slow inactivator character was indeed recessive (Table 9.1).

Dosage effect

Of great additional interest was the fact that a dosage effect was demonstrable. Thus the mean plasma isoniazid concentration of all rapid inactivators was found to be lower than that of a group of known heterozygotes. This is what would be expected if the homozygous rapids (who would be included under 'all rapid inactivators') had lower readings than the heterozygotes.

Table 9.1 Observed numbers of children of each phenotype compared with those expected on the hypothesis that slow inactivation of isoniazid is an autosomal recessive character. Remember that R can be heterozygous or homozygous. (From Evans *et al.*, 1960, by courtesy of the authors and the editor of the *British Medical Journal*.)

Expected numbers of children of each phenotype compared with those observed

Phenotypic matings	No. of matings	No. of children	R expected	R observed	S expected	S observed	χ^2	D.F.
S × S	17	54	nil	4	54	50	–	–
R × S	23	67	38.88	40	38.10	27	0.075	1
R × R	13	38	31.30	31	6.68	7	0.018	1
	53	159		75		84	0.093	2

$$P > 0.95$$

Anthropological aspects

From the anthropological point of view it is of interest that the proportions of the two phenotypes in an Indian population closely resemble that in the Caucasian and the American Negro. On the other hand, there is a much larger proportion (>90%) of rapid inactivators in the Japanese, and Eskimo populations consist almost entirely of this phenotype. The advantage of the dominant character is not known but it would appear that its possession is particularly helpful in the Far Eastern and Arctic environments.

Site of acetylation of isoniazid

Evans has shown conclusively by *in vitro* experiments that it is the rate of acetylation by an enzyme, acetyl transferase, that distinguishes the two classes and that it is in the liver that this takes place. He used fresh biopsy specimens of human liver from volunteers and after suitable preparation incubated them at 37°C with isoniazid. Two hours later the amount of free drug was estimated and from this the amount acetylated was calculated. The 'slow' and 'rapid' results obtained tallied with the inactivator status (determined later) of the volunteers.

Isoniazid as treatment

Generally speaking the slow acetylators are at more of a disadvantage because they are prone to toxic effects from accumulation of the non-metabolized drug molecules. The first example observed was polyneuritis in malnourished patients during treatment of their tuberculosis with isoniazid. In poor countries where tuberculosis is rife, investigations have been made to find the cheapest possible regimens which provide effective therapy. It was found that reducing the frequency of isoniazid dosage from twice to once weekly led to rapid acetylators having a significantly poorer response as regards their tuberculosis than slow acetylators. (Note that in the UK isoniazid is always given daily.)

A 'natural' metabolite

Many hundreds of enzyme polymorphisms are recognized but for very few of them is the natural substrate known. The so-called pharmacogenetic polymorphisms were obviously not designed by nature to deal with drugs which have only recently been the product of human inventiveness. The following new discovery is therefore of interest.

Caffeine is a plant component of tea, coffee, cola, chocolate and other edibles. The caffeine molecule (the substrate) has quite a complicated fate in the body. Amongst other compounds it produces 5-amino-6-amino 3-methyl uracil, which is subsequently acetylated in the 5 position. This acetylation step is polymorphic, controlled by the same genes as in isoniazid. It is in fact now possible to phenotype individuals by analyzing their urine after they have drunk a cup of coffee!

Relationship of the polymorphism to bladder cancer

The slow acetylator phenotype is associated with bladder cancer. This is of interest because the aromatic amines known to cause bladder cancer are polymorphically acetylated. The explanation for the association may be that slow acetylators are more susceptible to bladder cancer because they do not detoxify the causative amines as efficiently.

9.4 The hydroxylation polymorphisms

The most common fate of drug molecules in the body is for an oxygen atom to combine with hydrogen to form a hydroxyl group. This step takes place in the endothelial reticulum of the liver cells (which when the cells are broken up in the laboratory turns into 'microsomes' which can be isolated by centrifugation). The enzyme which mediates this step is known as P-450 (because it is a protein with a spectrophotometric absorption peak at 450 nm). Many endogenous chemicals are similarly metabolized. An example from the many recognized is now given.

The debrisoquine polymorphism

Debrisoquine is a post-ganglionic adrenergic (see glossary) blocker used to treat hypertension (high blood pressure). The drug was noted to be variable in its effects in different people, some individuals fainting on ordinary doses because of an excessive action in relaxing the arterioles. Investigation of the metabolism of the drug in the population was carried out by means of a simple test. Individuals swallowed one small dose and thereafter provided one urine sample. The urine was analysed to provide the 'metabolic ratio'.

$$\frac{\text{Concentration of unchanged debrisoquine}}{\text{Concentration of the metabolite 4 hydroxydebrisoquine}}$$

The frequency distribution of this ratio was bimodal with about 9% of the UK population sample being poor metabolizers (the 'fainters'). Investigation of pedigrees revealed the poor metabolizers to be Mendelian autosomal recessives.

Adverse reactions

The hydroxyl metabolite is eliminated quicker from the body than the original drug compound, and furthermore the hydroxyl metabolite is usually pharmacologically inactive. Because of these considerations people who are 'poor hydroxylators' are much more prone than 'extensive hydroxylators' to have adverse effects because they tend to have higher concentrations of the drug in their bodies after standard dosages.

The enzymic basis for the polymorphism

Pieces of liver were obtained from volunteers who were having abdominal operations. Subsequently the patients were typed for their debrisoquine status. It was found that poor hydroxylators had a virtual absence of a specific liver P-450 which oxidizes these drugs.

9.5 Induction and drug interactions

Phenobarbitone and some other drugs have the property of stimulating (inducting) the synthesis of enzymes, particularly those formed in the liver.

The drug warfarin is given to prevent further clotting of blood when thrombosis has occurred. Warfarin has this effect because it prevents the production of prothrombin, necessary for clot formation, from taking place in the liver. If prevention is allowed to proceed to an advanced degree the effect is lethal, and this is why warfarin is in use as a rat poison. Warfarin resistance is sometimes, both in man and rats, an inherited autosomal dominant trait. However, in patients who take phenobarbitone, which increases the metabolism of the drug (so that there is less available to inhibit prothrombin synthesis) resistance can also occur (see 'phenocopy' in glossary).

Phenobarbitone is also occasionally used in the treatment of haemolytic disease of the newborn. Thus it is sometimes given to an immunized mother a few days before delivery, or to the new-born baby, or both. Serum bilirubin (a bile pigment formed from degradation of haemoglobin from the red cells) concentration is then decreased in the baby, probably due to induction of enzymes in the liver, but its use is restricted to mild cases.

9.6 An interesting drug allergy

Almost everyone has been treated at some time with a penicillin preparation. In a large survey on twins it was found that amongst 432 monozygous twins (who had been exposed to penicillin) there was a concordance rate for adverse effects to penicillin of 34.8%. In the same survey amongst 382 dizygous twins similarly exposed the concordance rate for adverse effects to penicillin was 12.1%. The difference in concordance between the two sets of twins is statistically highly significant and indicates a powerful genetic influence on the liability to adverse effects from penicillin. Sensitivity to the drug is not uncommon, and generally the adverse effect follows re-exposure to it.

In the case of amoxycillin however, a rash not infrequently follows the consumption of oral capsules for the first time. This rash is often mistaken for rubella (German measles). The reason that this matters is that later on at school a girl may say that she has had German measles and therefore does not need vaccination. Never trust this story – always test for antibodies, for rubella in pregnancy may have disastrous effects on the baby.

Genetics or environment? often poses a problem but at first sight nowhere more so than in a pair of identical twins one of which was born with gross abnormalities, the result of maternal rubella, and the other perfectly normal. However, one has to remember that not infrequently one of a pair of identical twins is dead and perhaps there is here an even more striking difference between two genetically identical babies.

9.7 Alcohol

The definition of 'alcoholism' is more or less impossible and attempts produce marked differences of opinion. It is not difficult from family studies to show what might be single gene differences – 'alcoholism' or 'non-alcoholism' – the segregation appearing to be on Mendelian lines. However, these alleged clearcut differences depend on whose evidence is accepted. A quite sober husband may be taken to task by a fundamentalist wife for having a beer at a football match. Do you then score him as positive for 'family complains about his drinking habits'? In alcoholism, psychosocial habits seem to the writer to be much more important than genes and the Japanese produce good evidence of this. The Japanese are polymorphic for the genetically controlled aldehyde hydrogenase enzymes and about half the population have an unusual form which causes them to flush and have various other unpleasant vasomotor symptoms on consuming alcohol. It might be thought that these symptoms would encourage abstinence, but since the war the Japanese have adopted western style habits and drink far more than they did in spite of the side effects.

Still on alcohol, there is an interesting finding in a particular type of diabetes mellitus. Many cases require insulin, but a proportion can be managed on tablets called chlorpropamide. Some diabetics who are on the drug develop facial flushing when they take it, and it was originally thought that this was a dominantly inherited feature which could act as a marker for non-insulin dependent diabetic patients generally. Later work suggests, however, that the flushing only occurs in certain families; moreover, it is occasionally found in normal people who take alcohol without chlorpropamide. Nevertheless, diabetics who flush are less likely to develop retinal changes (often leading to blindness) than those who do not, this pointing to some genetic control of the eye complication.

10

Genetics and Preventive Medicine. The Prevention of Rhesus Haemolytic Disease of the Newborn (HDN)

10.1 The general problem

In Great Britain about 85% of the population, male and female, have on their red blood cells the Rhesus factor (a group of antigens) and they are then said to be Rhesus positive; the remaining 15% lack it and are Rhesus negative. The trait is inherited on normal Mendelian lines and for practical purposes Rhesus positive should be thought of as being dominant to Rhesus negative, though this is somewhat of an over-simplification.

If a Rhesus negative woman marries a Rhesus positive man she may conceive a Rhesus positive baby, the Rhesus factor having been inherited from the father. At or near delivery some of the baby's blood may leak across the placenta into the maternal circulation (see Fig. 10.1), in which case the mother may make antibodies to the Rhesus positive cells. However, the antibodies are slow to form, taking weeks or months, so that the first Rhesus positive baby is as a general rule unaffected, and the risk lies in subsequent pregnancies where the baby is Rhesus positive.

The clinical features of the disease are anaemia, oedema (swelling), jaundice and liver failure and occasionally Rhesus babies are mentally retarded or deaf. Sometimes − in fact often − the infant can be rescued by a complete exchange of all its blood (thereby getting rid of the antibodies) but mildly affected cases usually recover on their own. Nevertheless, the situation was a serious worry to Rh incompatible couples, since as many as 1 in 7 of all marriages are of this type.

In fact, however, the risk of a baby being affected has never been very high. For example, before prevention, when the total yearly birth rate was about 850 000, the number of 'Rhesus' babies was probably not more than about 5000 (with total deaths around 800) though particular families were very liable to have a succession of tragedies. Several natural factors operate to limit the risk of the mother forming antibodies. First, the Rh positive father may be hetero-zygous, having received an Rh negative gene from one of his parents, and there is then only a 50% chance that the baby will be Rh positive. Second, the leakage of the baby's blood through the placenta into the mother's circulation may not be in sufficient quantity to stimulate the production of Rh antibodies. Third, some women do not produce antibodies even if there is leakage. Fourth, in about 20% of cases the formation of Rhesus antibodies is prevented by the

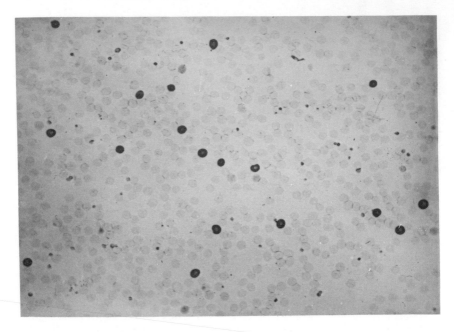

Fig. 10.1   Fetal cells in the maternal circulation. By means of a buffer the adult haemoglobin has been washed out (eluted) from the mother's red blood cells and these therefore appear as 'ghosts'. The fetal haemoglobin is not eluted by the buffer and the cells therefore stain darkly. This method of detecting fetal cells is known as the Kleihauer-Betke technique. (Redrawn from Clarke, 1964.)

protective mechanism arising from an interaction with the blood groups of another system, namely the ABO, where there are naturally occurring antibodies. The diagram (Fig. 10.2) shows how this protection operates. It was this phenomenon which gave the Liverpool group the idea of giving to the non immunized mother anti-Rhesus antibody (anti-D) after the delivery of a Rhesus-positive baby. The injection ('the Liverpool jab') is given into the mother's deltoid muscle and is virtually free from side effects. It eliminates any Rh-positive fetal cells that have earlier crossed into the mother's circulation and thus mimics the natural protection afforded by ABO incompatibility. The mother therefore starts her next pregnancy immunologically virgin, as it were, but she will need the 'jab' after each subsequent Rh-positive delivery.

   In retrospect, this seems a very simple procedure, but the research which led to it had its ups and downs and it may be of interest for readers to know some of the details.

10.2   Experiments on male volunteers

It was decided to test out the hypothesis on Rh negative men and in the first experiments we took a group of male volunteers and injected into them Rh posi-

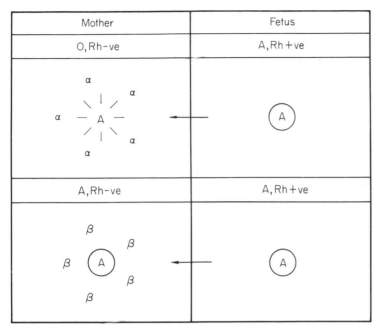

Mother	Fetus
O, Rh−ve	A, Rh+ve

| A, Rh−ve | A, Rh+ve |

Fig. 10.2 In the upper half of the figure a red cell from a group A Rh positive fetus is crossing the placenta and entering the circulation of a group O Rh negative mother where it will be immediately destroyed by the naturally occurring anti-A (α). Conversely, in the lower half, a red cell from a group A Rh positive fetus is entering the circulation of a group A Rh negative mother where it will survive normally, its life span being many weeks. NB Group O individuals possess naturally occurring anti-A and anti-B but the latter is irrelevant in this example and is omitted from the figure.

tive adult red blood cells labelled with radioactive chromium. Half of the volunteers were kept as controls and the other half were given the anti-Rh antibody about half an hour after the original injection.

The results were exciting − the injected anti-Rh did indeed get rid of a high proportion of the Rh positive cells. However, to our dismay we found that six months later, instead of preventing the men from forming antibodies, slightly *more* men had formed them than would have been expected by chance − we had possibly enhanced instead of suppressed antibody formation. However, we felt that our reasoning was basically sound, and very fortunately we did not let ourselves be discouraged. We tried a different kind of antibody and found that we had given the wrong type the first time. We had given the 'complete' form of antibody, which acts in saline solution, and this we found had left the residue of the cells still antigenic even though the cells themselves had gone. We therefore did a second set of experiments using 'incomplete' anti-Rh, which 'coats' the antigen so that it cannot make contact with the antibody-forming cells. These were much more successful and prevented antibody formation in almost all the subjects treated.

10.3   Experiments with fetal blood in women volunteers

This same procedure was found to be successful when *fetal* (instead of adult) red cells were used, and when the subjects were Rh negative *women* volunteers who were past the child-bearing age.

10.4   Relation of the formation of Rhesus antibodies to the size of the feto-maternal bleed

By means of a survey among Rh negative women, we also established that the likelihood of their forming antibodies against their Rh positive children generally depended directly on the *number* of fetal cells which had crossed the placenta, and these could be recognized by the Kleihauer-Betke technique (see Fig. 10.1).

10.5   The Liverpool clinical trial

We felt, therefore, that the time was ripe for trying the treatment on primiparae (i.e., women delivered of their first baby) who had had a Rh positive, ABO compatible baby and had had a considerable transplacental haemorrhage since these women would be the ones most seriously at risk. The importance of the choice of this 'high risk' group should be clearly realized – it is not necessary to know much in the way of statistics to appreciate that if we treated these women (and kept similar women as controls) we should know much sooner whether the treatment was successful or not than if we treated women who were less at risk. The experiment was carefully planned to give an answer *quickly*, before many women had been subjected to a procedure which might be of no use and might even have unforeseen disadvantages or dangers.

We therefore took from five Liverpool maternity hospitals all Rh negative women who had had a Rh positive, ABO compatible baby and who had had a 'big' bleed, i.e. 5 cells or more of fetal blood per 50 low power fields of the microscope, and we treated *alternate* ones, the others being kept as controls. The treatment consisted in giving 5 ml (about 1000 micrograms) of anti-Rh obtained from our volunteers or previously immunized women. It was given, in the form of gammaglobulin, to the mother within 36 hours of delivery, and we tested to see if the fetal cells were in fact diminished during the next few days, and after six months – when the mother might have begun to make antibodies and the injected antibody would almost certainly always have disappeared – to see whether the mother had made anti-Rhesus antibodies. The results were astonishingly successful; practically no treated mothers made antibody and many controls did; it was far better than we had hoped; in all biological procedures there is variation and error and we should have been perfectly satisfied if we had protected 75% of cases. In fact we protected far more – about 95% – and later a much smaller dose (200 micrograms) was found to be equally effective.

10.6   Results in England and Wales

Anti-D prophylaxis was introduced via the NHS about 1970 so that at the present time we have had around fifteen years experience of it. Figure 10.3 shows the number of liveborn deaths and it will be seen that although the numbers were falling since 1950 the decline was much accelerated following 'the jab'. Some of the improvement might have been due to better obstetric care but the fact that the number of failures and those cases where the woman was immunized during pregnancy did not fall substantially between 1977 and 1983, suggests that improved treatment of haemolytic disease, as contrasted with improved prevention, did not play a major part in the overall decline in mortality from the disease during this period. The graph for stillbirths from Rhesus disease gives much the same results but from now on we think that the distinction between these two notifiable causes of death should be abandoned. This is because if a Caesarean section is done say at 27 weeks and the baby draws a few breaths and then dies it receives a death certificate. If, however, it is delivered dead at this time it does not rank as a stillbirth since these are not recorded until after the 28th week of gestation – it is in fact not identified at all.

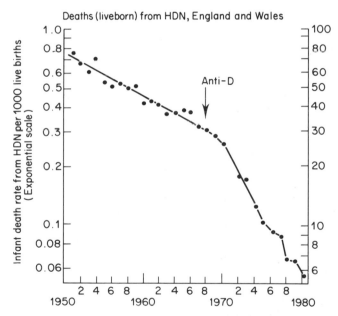

Deaths (liveborn) from HDN, England and Wales

Fig. 10.3   Graph showing accelerated decline in liveborn deaths in England and Wales from Rhesus haemolytic disease since the introduction of prophylactic anti-D gamma-globulin. Figures on the right are deaths from HDN per 100 000 live births. If the birth rate is around 600 000, multiplying the figures on the right by 6 gives the approximate number of actual deaths. Our estimates for total deaths (live and still born) from Rhesus (D) HDN in recent years: 1977, 106; 1978, 90; 1979, 87; 1980, 72; 1981, 41; 1982, 44; 1983, 34; 1984, 24; 1985, 33.

The fact that deaths from Rhesus disease have not been entirely eliminated is due to several causes.

(*a*) Even in 1986 there are still a few women having babies who were immunized before prophylaxis was available.

(*b*) A considerable number of patients still do not receive the 'jab' even when it is indicated; this is particularly so following an abortion when the patient often leaves hospital very quickly and the injection is forgotten.

(*c*) A few women become immunized during the last three months of their pregnancy and if this is the case the 'jab' given after delivery is too late. This has been realized for some time and in Canada and Australia antenatal anti-D as well as postnatal is the rule and this has reduced the incidence of the disease very considerably. However it means that a great deal more gammaglobulin is needed because normally the baby cannot be grouped before delivery and therefore the injection may be given unnecessarily when the baby is Rhesus-negative.

(*d*) There is a genuine small failure rate, no medical procedure ever being a hundred per cent effective.

Dr L. A. D. Tovey, Director of the Blood Transfusion Centre in Yorkshire, has been interested both in deaths from Rhesus disease and in the immunization rate year by year in Rhesus-negative mothers, and Fig. 10.4 shows what is happening in his region compared with that in Manitoba and Connecticut. Tovey has also carried out an excellent antenatal anti-D trial in Yorkshire. A 100 microgram dose of anti-D immunoglobulin was given at 28 and 34 weeks to 2069 Rh(D) negative primigravidae as well as post partum women. The reason that first pregnancies were chosen was that they were unlikely to have been sensibilized to the Rh antigen before the antenatal injection and thus the effect of the antenatal prophylaxis could be more easily assessed. For comparison

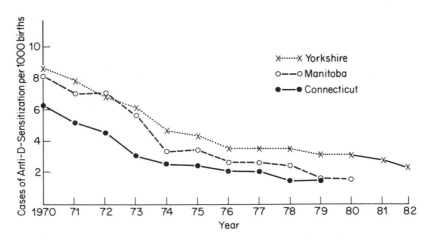

Fig. 10.4   Incidence of maternal anti-D sensitization per 1000 births reported by three centres. (Redrawn from Tovey, 1984. By courtesy of the author.)

2000 primigravidae delivered of an Rh(D) positive infant who only received the standard post delivery injection were used as controls.

The results showed that 2 of the 2069 in the trial group and 18 in the control group developed antibodies in their first pregnancy. In subsequent Rh(D) positive pregnancies, 2 of the trial group and 22 of the control group developed antibodies. These results are statistically significant

$$\chi_1^2 = \frac{(O-E)^2}{E} = \frac{(2-18)^2}{18} = 14, p < 0.01$$

10.7 A theoretical consideration: the specificity of anti-D

Thinking superficially about the prophylaxis by anti-D, one might assume that since only the D antigen sites were blocked, the other antigenic determinants on the red cells would be quite capable of stimulating the appropriate antibody. However, there is the possibility that coating a single antigen site damages the cell so that all determinants are suppressed. This seems to be the case when there is ABO incompatibility between mother and fetus, Rh immunization being suppressed here. The Liverpool 'Kell' experiment was designed to test the two hypotheses, and it was found that giving anti-Kell (Kell is another red cell blood group system) to Kell negative, Rh negative men challenged with Kell positive, Rh positive cells, prevented not only anti-Kell from being produced but also anti-Rh antibody. So non-specificity was shown to be true, anyhow for these two systems, and the principle might have applications in the prevention of rejection in transplantation. For example, antibodies directed at any determinant on the donor cells should suppress the immune response to other determinants present.

10.8 Plasmapheresis

In women who are immunized it is possible to lower the anti-Rh antibody titre by repeated removal of the plasma, the red cells being replaced. The trouble is that the antibody forming cells quickly start working overtime and the titre rapidly comes back to what it was before the procedure. However, there are some cases which seem to have benefitted and it is popular with patients as they can understand what is being done. However no proper control trial has ever been carried out and each individual case is a law unto itself. Personally we are doubtful as to the efficacy of the procedure but many would disagree with this view.

10.9 Oral antigen

By stripping the red cells of their membrane, it is possible to make a concentrated extract of the Rhesus antigen and there is some evidence that when this is made into capsules which are given orally to immunized women the plasma level of IgG anti-D is lowered because it is converted into IgA and IgM anti-D; if this is so neither of these will cross the placenta and affect the baby. The two names associated with this form of treatment are Bierme (France) and Beer (USA) and work is still proceeding particularly to see whether the treatment also prevents primary immunization. (Barnes *et al.*, *Clinical and Experimental Immunology* (in press) find no evidence to support this suggestion.)

10.10   Haemolytic disease of the newborn foal

The blood group systems of the horse are quite different from those in Man, but haemolytic disease of the newborn foal is produced essentially by a similar incompatibility mechanism to that in Man. However, the antibody formed by the mare does not cross the placenta, but reaches the foal via the early milk (colostrum) and it is highly toxic for 36 hours. Provided it is known that the mare is immunized, the foal can be prevented from suckling and the disease is not a problem.

Racehorses are particularly liable to this disease, Shire horses less so and ponies almost never, the risks being related to the absence or presence of naturally occurring antibodies.

What happens in the mule (the cross between a mare and a jack donkey)? One would think that every mule after the first would be affected but how many mares have more than one mule? The books say that 8% of all mules are affected but how this figure is arrived at is not stated – a good topic for research in departments of veterinary science.

Key papers dealing with the discovery of the Rh blood groups, the management of haemolytic disease of the newborn and the development of the prophylaxis are reprinted with commentaries in Clarke (1975). The 'jab' is a good example of how one may outwit one's inheritance.

11

Genetic Advice to Patients

11.1 General

With the spread of medical knowledge there is an increasing desire to read the
future and the commonest request is for information about risks to subsequent
children when one, quite unexpectedly, has been born with a defect, to normal
parents. Alternatively, individuals about to get married may seek advice about
risks to offspring, because of a bad family history, or perhaps because they are
marrying their first cousins. Thirdly, parents knowing of a skeleton in the
family cupboard may ask about the chances of it turning up in their grandchildren – though it should be remembered that not infrequently what they
really want is medical backing against a union of their children they dislike on
other grounds.

Advice on genetic matters should always be given in terms of probability,
never certainty, and in these days of football pools it is generally assumed that
patients readily understand odds. On reflection however it is possible to take
the opposite view that no one who really understands odds would attempt football pools.

The basic information for patients is that about one pregnancy in thirty will
produce a baby with an abnormality which appears either at birth or very early
in life e.g. harelip, spina bifida, congenital heart disease or mental defect.
'Congenital' simply means recognizable at birth and does not necessarily mean
that the condition is inherited.

11.2 Disorders where fairly precise information can be given

This is only possible in a minority of cases i.e. those which show clear-cut
Mendelian inheritance and when this is the case the risks to subsequent offspring are usually unacceptable to most people, assuming that the disease is a
serious one.

Examples
(a) In a disorder controlled by an autosomal dominant gene e.g.
Huntington's chorea (see p. 1) the chance of any given offspring having the

disease if one parent be affected is 1 in 2 and the risk is similar for subsequent siblings − chance has no memory.

(*b*) If a child is born with a recessive trait, e.g. fibrocystic disease (see p. 3), the risk of any subsequent offspring being affected is 1 in 4. The same is true of phenylketonuria though there are pitfalls (see p. 5).

(*c*) In the Down syndrome, when it is caused by a translocation (see p. 62), there is *less* than the expected 1 in 4 chance that the child of a carrier will be affected.

(*d*) If the disease is due to an X-linked recessive gene, e.g. haemophilia, or the fragile X syndrome, then an affected male married to a normal woman will have all carrier daughters but all his sons will be unaffected. On the other hand, of the daughters of a carrier female half will be normal and half carriers and of her sons half will be affected and half normal. In the case of the sex-linked Duchenne muscular dystrophy the boys are practically always infertile. If in the above conditions there is no family history it is very important not to give the rather bad prognosis for relatives until one has considered the very rare possibility of a mutation. If this had occurred in embryonic life to produce a haemophilia patient then none of his sisters will be carriers. It may be very difficult to decide the point but enquiries about the disease in maternal uncles and great-uncles are important.

11.3   Diseases where the risk is low and empirical

Here, though there is a genetic component to the illness, it is not transmitted in any clear-cut way. This may be for various reasons:

(*a*) the disorder may be determined by many genes (polygenic inheritance) e.g. hypertrophic pyloric stenosis, congenital dislocation of the hip (in part) and probably high blood pressure;

(*b*) the environment may be also partly responsible (multifactorial inheritance) e.g. anencephaly and spina bifida (ASB);

(*c*) the disease may be heterogeneous with differing factors responsible in the various subgroups e.g. coeliac disease, a diarrhoeal disease due to gluten sensitivity; here there is an association with the HLA groups but this varies between families.

The definition of empirical is 'the probability of occurrence of a specified event based upon prior experience and observation rather than on prediction by general theory' (Herndon, 1962).

Empirical risks to future children (an earlier one having been affected) are usually small enough to be acceptable to most parents. A rough guide in these cases is that the rate of occurrence is approximately equal to the square root of the prevalence in the general population. Since the prevalence of most disorders where the genetic component is ill-defined is between 1 in 600 and 1 in 2500, the recurrence rate would be between 1 in 25 and 1 in 50, e.g. in congenital dislocation of the hip about 1 in 25; in hypertrophic pyloric stenosis 1 in 50 if the index case is a male and 1 in 10 if female (see also p. 16); in anencephaly and spina bifida 1 in 30. Roberts and Pembrey (1978) make the good point that if one has to do many calculations to arrive at an empirical risk factor the answer is usually acceptable to the parents and therefore the working out has been somewhat of a waste of time.

11.4 Notes on some disorders where the risks are empirical

Congenital dislocation of the hip

Here infants are born with the disorder, which varies from a minor radiological change in the acetabulum (the socket in the pelvis for the hipbone) to a severely incapacitating lesion. Several factors have to be taken into account. Two of these are genetic, one affecting the development of the acetabulum (probably multifactorially controlled) and the other producing an increased laxity of all joints (a dominant trait). There is also an environmental component, namely the intra-uterine posture. At birth a clicking hip has been thought to be an early sign of the disease but my colleagues tell me that any good orthopaedic surgeon can make a baby's hip click.

Hypertrophic pyloric stenosis (see also p. 16)

Here there is pathological hypertrophy of the sphincter between the stomach and jejunum. The condition occurs in infants (particularly boys) and leads to intractable vomiting because of the obstruction. A simple operation whereby the hypertrophied muscle fibres are split usually gives good results. An interesting observation is that affected children subsequently have exceptionally well developed musculature.

Anencephaly and spina bifida (ASB)

These two related neural tube abnormalities receive special mention since they are fairly common and involve both nature and nurture components. Furthermore they cause a great deal of parental suffering and their management sometimes raises difficult ethical problems. In anencephaly a part of the skull and brain fails to develop and the condition is incompatible with survival for more than a few days. In spina bifida there is a defect in the lumbo-sacral region of the vertebral column (i.e. usually below where the actual spinal cord ends) and the nerve roots and their coverings may protrude. Although sometimes the condition is trivial in others the roots are damaged and weakness or paralysis of the lower extremities occur. There may also be sphincter incompetence. The incidence of ASB varies greatly even within quite small geographical areas and it is probable that a high proportion of affected fetuses are lost as spontaneous abortions. It has even been suggested that geographical variations may be due more to the proportion aborted than to a true difference in incidence and conception. The highest incidence in the world is in the big cities of Ireland where as many as 9 per 1000 of live births are affected and ASB is generally commoner in social classes 3, 4 and 5 than in 1 and 2.

The cause of ASB remains unknown. Environmental factors, e.g. potato blight, have been suggested and so have genetic influences but the problem remains unsolved. It is possible that the micro-environment of the uterus is abnormal at the time of conception of the malformed baby, since when the mother has had a spontaneous abortion, it is twice as likely to have occurred in the pregnancy immediately before, than in the one immediately after, the ASB baby. This finding only relates to women who had had a pregnancy on either side of an ASB baby.

An attempt at the prevention of ASB must take into account the fact that the neural tube closes between the 3rd and 4th week of pregnancy so that often the condition is established before the mother knows she is pregnant. This led Smithells and his colleagues (1981) to suggest treatment before conception by supplementing the diet with a compound vitamin pill (containing folic acid) and this has given some promising results. However it must be borne in mind that ASB is much commoner in this country than in the slums of Calcutta and simple vitamin deficiency is therefore unlikely to be the whole cause. Similarly in South Africa ASB is more prevalent in whites than in blacks or coloureds (Cornell *et al.*, 1983).

Diabetes mellitus may hold a clue, for mothers with this complaint are much more likely to have babies with congenital malformations (including ASB) than are non-diabetic women. However, if the diabetes is properly controlled from a very early stage of pregnancy, there is some evidence that the malformation rate is then the same as that in non-diabetic women. This may mean that conditions such as ketosis or hyperglycaemia are relevant to ASB (and other malformations) and it may be that excess ketone bodies in normal women were a factor and that vitamin supplementation corrects this – a hypothesis only, but testable.

The discovery that the levels of alpha-feto protein (afp) in the amniotic fluid are raised in open neural tube defects means that in anencephaly almost all cases are detectable before birth, as are 95% of open spina bifida (see also ultrasound p. 95). Interestingly, in Down's syndrome the amniotic afp level tends to be low, possibly because the disorder is characterized by retardation of growth in almost every part of the body (Lionel Penrose's observation). In practice, estimating the afp level is not a very reliable test in Down's syndrome, and again the possibility of causing the abortion of a normal conceptus by amniocentesis must be borne in mind. This method of detection is useful if the mother has already had a deformed child, the risk of recurrence being about 3%, but amniocentesis (see p. 94) is not without hazard and carrying it out in all pregnancies is certainly not practicable. It is possible using a radio immunoassay technique (see glossary) to test the mother's serum for alpha-feto protein and with this technique about 90% of anencephalics and about 70% of open spina bifidas can be detected. Having collected this information, however, ethical problems are immediately raised. In this context, anencephalics need not worry us as they always die in a few days and, within a few months, so do most of the spina bifida babies with very large defects. In those cases where the defect is smaller it is impossible to know from the test whether or not babies will be mentally affected and this to many people is crucial. Should a woman be advised to have a therapeutic abortion of a fetus that might be mentally normal even though paralysed? Seeing them play in their special schools and hearing the favourable remarks of many parents makes one hesitate. Prevention would solve the problem.

Disseminated sclerosis (multiple sclerosis)

Here there are scattered areas of scarring in the nervous system, the result of localized destruction of the myelin sheath which protects the white matter of the brain. The symptoms are extremely variable but transient blindness, weakness of the legs and numbness in the feet are often present. It is the commonest organic disease of the central nervous system in the United Kingdom, the prevalence being about 50 per 100 000. A good general rule is that the risk rises with increasing distance from the Equator although there is an exception in Japan where the disease is unexpectedly relatively infrequent. In South Africa immigrants from temperate climates carry their risk with them. The chance of developing the disease is approximately 15 times greater in the first degree relatives of patients than in the general population. However no clear pattern of transmission has emerged, but some support for the influence of genetic factors derives from the significantly increased incidence of the histocompatibility antigen HLA3 in patients with the disease. In general, however, unknown but non-genetic factors are thought to be responsible for the disorder.

11.5   Detection of the carrier state

This is particularly important in the case of X-linked conditions, for example in advising the sister of a man with haemophilia since if she is a carrier half her sons will be affected, and Table 11.1 gives some X-linked abnormalities which can be tested for in women who have affected relatives.

Table 11.1   Carrier detection in X-linked disorders i.e. in women who have an affected relative

Disorder	Abnormality
Haemophilia A	factor VIII reduced*
Haemophilia B	factor IX reduced
G6PD deficiency	erythrocyte G-6-PD reduced
Lesch-Nyhan syndrome[1]	hypoxanthine-guanine phosphoribosyl transferase in skin fibroblasts reduced. Two populations of cells
Hunter's syndrome[2]	iduronosulphate sulphatase reduced in skin fibroblasts. Two populations of cells
Ocular albinism	patchy depigmentation of retina and iris
Vit. D resistant rickets (hypophosphataemia)[3]	serum phosphorus reduced
Duchenne muscular dystrophy[4]	serum creatine kinase raised

Table 11.1 *Cont'd*

Disorder	Abnormality
Becker muscular dystrophy[5]	serum creatine kinase raised
Diabetes insipidus (nephrogenic)[6]	urine concentration diminished
Fabry's disease (angiokeratoma)[7]	α-galactosidase in skin fibroblasts reduced. Two populations of cells
Retinitis pigmentosa[8]	tapetal reflex
Anhidrotic ectodermal dysplasia[9]	sweat pore counts reduced, dental anomalies
Lowe's syndrome[10]	lenticular opacities

* More precisely a reduction in the ratio of factor VIII activity to inactive antigen.

Key to disorders
1. Characterized by mental retardation, self-mutilation and neurological abnormalities.
2. One form of mucopolysaccharidosis.
3. A particular form of rickets not due to vitamin D deficiency.
4. This is the severe form.
5. This is the mild form.
6. Characterized by the excretion of large volumes of dilute urine due to abnormalities in the kidney.
7. A disease of the skin and blood vessels.
8. Characterized by retinal atrophy, clumping of the pigment and night blindness. There are other inherited forms but the tapetal reflex (a sheen on the retina) is only found in the sex-linked form.
9. Inability to sweat. Absent or defective teeth and nails.
10. Characterized by cerebral, eye and kidney symptoms.

In autosomal recessive conditions carrier detection has little application except in the case of cousin marriages where there is a relevant family history. (In healthy families there is no contraindication to cousin marriage, but in certain populations the matter is important – for example, the Ashkènazi Jews are very liable to Tay-Sachs disease, a severe autosomal recessive disorder which produces blindness, mental deficiency and death in early childhood. The disease is now much less common since members of the sect have heeded the general advice against cousin marriage.) On the other hand, screening for the homozygote, for example in phenylketonuria (see p. 5), is very important, and where a disease is common, e.g. thalassaemia, screening a population, as has been carried out in Italy, is useful in detecting carriers who wish to marry.

11.6   Aids to prenatal diagnosis in the fetus

Amniocentesis

The amnion is a fluid-filled sac in which the embryo develops and by means of a hollow needle 5–10 ml of fluid can be removed through the abdominal wall. The cells in it are fetal and they can be sexed by the Barr body (see p. 58) or grown and examined for chromosomal or biochemical abnormalities if there is reason to suppose an affected baby might be born. (Sexing is essential for sex-linked conditions.) Furthermore, sometimes the fluid contains a protein (alpha feto-protein) from the fetus and when a high level of this is found it is likely that it has leaked from a neural tube lesion though other causes may give a positive result.

Knowledge obtained from amniocentesis is very helpful in deciding on the management of the pregnancy including possible termination, but as already stated, it is not entirely free from risk and abortion may be precipitated (variously estimated from 1 to 2%) so that it should not be done as a routine test.

More recent procedures are ultrasound, fetoscopy, fetal blood sampling and various techniques of molecular genetics (see Chapter 12).

Ultrasound

This is based on the principle of transmitting high frequency low intensity ultrasonic waves through the body. The reflected waves map the shapes and boundaries of organs and the technique is non-invasive and entirely safe. It enables the obstetrician to locate the placenta, establish the gestational age, diagnose twins and confirm fetal death. It can also usually recognize anencephaly and some cases of spina bifida – and congenital heart disease is now on the list. A most important point in assessing the accuracy of ultrasound results is the experience of the operator, but in expert hands the technique is tending to supplant amniocentesis.

Fetoscopy

This is carried out by a flexible instrument which can be inserted into the amniotic cavity under local anaesthesia so that the fetus can be directly inspected. It is of particular use if malformations of the limbs, face or genitals are suspected but it is not without risk and is more liable to produce an abortion than is amniocentesis.

Fetal blood sampling

Here the procedure is to remove under direct vision fetal blood from an umbilical vessel so that in the case of Rhesus disease the blood group of the baby, its haemoglobin percentage, the Coombs' status (see glossary) and the bilirubin level can be established. Fetoscopy can be done between 18 and 20 weeks and is carried out at the same time as blood transfusion is given via the vein, particularly in cases where the baby is hydropic. This direct intravascular blood transfusion is becoming increasingly used since it can be given sufficiently early to save the baby, whereas the standard procedure of intraperitoneal intra-uterine transfusion cannot be done until later. Fetal red cell sampling is also of great value in diagnosing thalassaemia in a fetus whose parents are both heterozygous for the condition. Thalassaemia is a heterozygous disease which occurs mainly in the Mediterranean area and in Asia. In parts of Italy over 10% of the population are carriers of the α-thalassaemia gene and in the homozygous state it causes a severe haemolytic anaemia which is usually fatal in childhood.

Obtaining a sample of fetal *serum* (as opposed to cells) is very important in the prenatal diagnosis of haemophilia, where factor VIII activity (see p. 8) is absent in the fetal serum if the (male) baby is affected.

Restriction fragment length polymorphisms

The use of this technique is explained in Chapter 12.

12

Molecular Genetics

12.1 DNA

In 1869 a German physician called Miescher made a chemical analysis of the nuclei of human pus cells, collected from bandages. The extracts were acidic (hence 'nucleic acids') and were found, using standard methods of the analytic chemist, to be rich in nitrogen, sulphur and phosphorus. Over the years, it was found that the two main kinds of nucleic acids were deoxyribose nucleic acid (DNA) and ribonucleic acid (RNA), but there was little knowledge about their roles in living organisms until about the middle of the present century. Then Avery, in 1943, found that the non-pathogenic R strain of the bacterium *Pneumococcus* could be converted to the S strain (pathogenic) by simply adding DNA from the S strain to the medium in which the R strain was growing. However, the important point was that the pathogenicity was permanent and was passed to subsequent generations when the bacterium reproduced. Therefore, here was a convincing piece of experimental evidence to suggest that DNA is the substance responsible for the characteristics of living organisms.

Though the chemical constitution of DNA is important, much more so is the arrangement of the component parts of the molecule; for it must be remembered we are dealing with the chemical which is the origin of all life, whether bacteria, fungi, plants or animals and there is excellent evidence that all living creatures evolved from this common ancestor. In 1953, Watson and Crick, using a technique known as X-ray diffraction, originally employed for giving information about the arrangement in space of the atoms within the molecules of crystals, proposed for DNA the well known double helix, resembling a spiral staircase. The reason for their proposal was that repeats in the pattern came at much longer intervals than those found when the DNA was examined chemically. This could be explained if the X-rays were only revealing that proportion of links which were seen from the same angle, a result which would be expected if the chain were coiled in a helix. The details of the helix are shown in Figs 12.1 and 12.2 (pp. 98 and 99) and it is now important to summarize the necessary properties of heritable material; it must be stable, able to replicate, to hand on information and to mutate. DNA fulfils all these criteria. Thus:

(a) It will stand boiling for one hour.
(b) Replication can be effected. The generally accepted explanation is that if

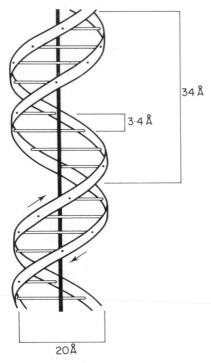

34 Å

3·4 Å

20 Å

Fig. 12.1   The double helix of DNA, giving dimensions in Ångstrom units (a unit of length equal to 10^{-1} m). The two ribbons symbolise the two phosphate-sugar chains and the horizontal rods the paths of the bases holding the chains together. The arrows show that the sequence of bases goes one way in one chain and the opposite way in the other. The vertical line represents the axis of the molecule. (Redrawn from Clarke, 1964.)

the two chains which are wound round each other separate, each single chain will have a series of bases needing new partners. These are taken from the nucleotide pool (DNA can only replicate itself in a living cell) and the appropriate freshly synthesized nucleotide bonds itself to the right base on the old chain (guanine plus its sugar and phosphate if cytosine is needing a new partner, adenine plus its sugar and phosphate if thymine is needing one, and so on).

(*c*) Information can be transmitted and varied by altering the order in which the pairs of bases are found. For instance

G	C
C	G
A	T

would carry a different message from that conveyed by

A	T
G	C
T	A

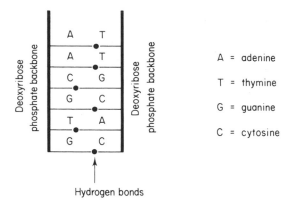

Fig. 12.2 The nucleotides, that is the units of DNA, when paired, have a backbone of the pentose-phosphate chain, and the nitrogenous bases (the nucleosides) are directed inwards and joined together by hydrogen bonds. Adenine (a purine) always pairs with thymine (a pyrimidine) and guanine (a purine) with cytosine (a pyrimidine). (Redrawn from Clarke, 1964.)

NB. The bases can follow one another in any order and the ratio of adenine or thymine to guanine or cytosine is different in different species. The ratio of total purine to total pyrimidine is, however, always unity.

Also necessary for the transmission of information is ribonucleic acid, which bridges the gap between the DNA in the chromosomes and the amino acids in the cytoplasm. There are three different forms of RNA:

(*i*) messenger RNA (mRNA) which conveys from the nucleus of the cell to the ribosomes (small particles of nucleic acid) in the cytoplasm the necessary information to ensure that a specific protein molecule is actually made;

(*ii*) transfer RNA (tRNA) which carries individual amino acids from cytoplasm to ribosome;

(*iii*) the RNA in the ribosome itself.

(*d*) Mutation can occur by mistakes in pairing of bases.

It was over twenty years before the molecular structure of DNA made an impact on medicine and what has happened in the last decade has opened up an entirely new era, which my one-time house-physician, now Professor D. J. Weatherall, has aptly called 'the new genetics' (see Weatherall, 1986).

From what has been said it might be thought that DNA/RNA is the universal basis of life. However, scrapie, the neurological disease of sheep and goats, is caused by an organism containing neither DNA nor RNA and called a prion, posing yet further problems in evolution.

12.2 The new genetics

The essence of this is 'genetic engineering' – but the more scientific term is 'molecular genetics' because the techniques are used at the DNA and not the cellular level.

The essential feature of this new discipline is that DNA fragments from two different organisms can be combined (technically but erroneously referred to as '*re*combined') to produce functioning hybrid DNA molecules. In this way, one is bypassing the barriers normally imposed by sexual incompatibility. To do this the 'donor' DNA, for example from a patient, is incorporated with the 'recipient' DNA of a virus (a bacteriophage) which infects bacteria. This hybrid DNA is then inserted into a bacterium, usually *E. coli*, so that replication can take place and many copies of the hybrid DNA can be produced. It is the 'many copies' which distinguish hybrid DNA from what happens in naturally occurring species crosses such as the mule and the hinny.

Given this, one might think, at the medical level, it would be possible to exchange a bad gene e.g. that causing phenylketonuria for a good i.e. normal one, but in practice in Man this is decades away, for three principal reasons.

(*a*) The inserted gene might not be correctly sited in the appropriate chromosome, nor be under the same control as the home product.

(*b*) The insertion might be carried out at the wrong stage of development and some phenylketonuric genes might be left behind.

(*c*) The vector (the virus) might cause chromosomal damage.

The mind boggles at the problems – both technical and ethical. However, transfection (see glossary) is now with us in animals and plants. Where therefore lies the present clinical use of recombination? The answer is in greatly refining prenatal diagnosis and the detection of carriers (neither of which usually present serious ethical problems. (See Emery, 1984, for an introduction to recombinant DNA.)

12.3   Prenatal diagnosis and the detection of carriers

The aim is to find that section of a strand of DNA or sometimes RNA which carries the gene responsible for a particular genetic disease. We therefore start with a patient with that disease and extract the DNA from the lymphocytes, a standard procedure. This is then used as a probe to test the DNA of an individual at risk, often a relative, to find out whether he/she has the disease, is a carrier or is free of it. Where the DNA sequence of the actual gene is not known, it is sometimes possible to find other markers such as, in the old days, a blood group, which are closely linked to the disease producing gene. When this has been done it may be possible to determine who in the family has the faulty gene and who has not. The procedures used are discussed next.

Restriction enzymes (restriction endonucleases)

Many bacteria make enzymes called restriction nucleases which protect them by cutting up any invading foreign DNA molecules, such as those present in viruses. Each enzyme recognizes a specific sequence of from four to six nucleotides in DNA from any source, and it always cuts at the same site, though different enzymes cut at different sites. The bacterium is prevented from destroying its own DNA because this is camouflaged chemically, but any foreign DNA which enters is immediately recognized and attacked. The restriction enzymes are so called because their activities are restricted to destroying foreign DNA.

Restriction nucleases have been purified from many species of bacteria and more than 100, most of which recognize different nucleotide sequences, are now commercially available. Their nomenclature derives from the various species of bacteria from which they are obtained. Thus HpaI is from *Haemophilus parainfluenzae*; Eco RI, *Escherichia coli*, and Hind III from *Haemophilus influenzae*.

Figure 12.3 shows the DNA sequences recognized by three commonly used restriction nucleases and it will be seen that the cut can be either straight or staggered; the relevance of this is important in the joining up of the introduced DNA molecules (from the patient) with those of the viral vector.

As has already been said the use of the nucleases is to help in identifying genes responsible for diseases or markers close to them. In this connection, particularly with markers, it is important to appreciate the fact that the total restriction site pattern is different in different individuals (except for identical twins) like the total blood group complement. This is because of variations in nucleotide sequences, though these differences, like the blood groups, have no

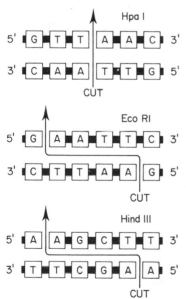

Fig. 12.3 Three DNA sequences recognized by HpaI, EcoRI and Hind III respectively. Such sequences are often, as in these examples, six base pairs long and palindromic. When the cleavage is staggered the ends are 'sticky' and easier to join with the enzyme DNA ligase than they are when the cleavage is blunt. (Adapted from Alberts *et al.* (1983), by permission.)

apparent phenotypic effect. Again, like the blood groups, they are inherited in a Mendelian manner, though there is no dominance, and restriction fragment length polymorphisms (RFLPs) result.

In a family segregating for different lengths of restriction fragments one can observe which length is possessed by the patient with the disease and which by 'normals'. An individual who appears to be normal but who has the 'disease' restriction fragment is a carrier and may develop the disease or hand it on. As has already been said, there is no dominance in the RFLs so that the possession of a certain allele cannot be masked by the one on the other chromosome, and this fact is of course very useful when one is studying a pedigree. It is also useful to know on which chromosome the gene being looked for is situated and this can sometimes be found by the cell fusion technique described on p. 68. Having identified the chromosome a restriction map can be constructed which shows the location of each cutting site in relation to its neighbours and this again obviously helps in studying pedigrees.

An example of a disease where the RFLP technique has been very valuable is Huntington's chorea (p. 1) for the disease gene is linked to a polymorphic DNA marker on chromosome 4. In Duchenne dystrophy (X chromosome), in phenyl-ketonuria (chromosome 12), and now in cystic fibrosis (chromosome 7), there are similar advances. However, it must be remembered that all these are applicable only in certain families where the RFLPs segregate appropriately – and crossing-over also has to be taken into account. Furthermore, linkage with RFLPs tells us nothing about the nature of the gene (any more than does the ABO blood group about the nail-patella syndrome (see p. 37)). It might be thought that with modern computer technology a study of the RFLP lengths by themselves would make it possible to say which particular RFL was associated with the disease but this is not so, and in all cases searches for the gene or marker a probe is needed. This is discussed next.

Gene probes

A gene probe is a length of single-stranded DNA with a specific nucleotide sequence, used to seek out and find complementary sequences in the DNA being investigated, i.e. in a possible sufferer from, or carrier of, a genetic disorder. In the case of the haemoglobinopathies these sequences are well known and the probe is therefore accurate. Sometimes, in fact often, the probe is much less certain, the sequences being those of the restriction length which is close to the gene (see above). Alternatively, a promising DNA sequence may be obtained from a genomic library where the entire genome of a particular tissue is stored. The matter is extremely complicated and there is often an element of guesswork and luck in the probe which is selected. Here we give the classical and easily understood example of 'seeking and finding' for thalassaemia, an inherited haemoglobin disorder, common in Italy, Greece and parts of the Middle East and Asia.

We start with the messenger RNA (mRNA) obtained from reticulocytes, which are the nucleated precursors of the red blood cells where the thalassaemic gene will certainly express itself. In order to make the probe, the mRNA has first to be changed into complementary DNA, (cDNA) and this can be affected by treating it with reverse transcriptase. This cDNA is made radioactive (so that it can be seen) and when we have got rid of the mRNA by means of alkali we have a single-stranded probe which will seek out and find a complementary sequence among the cloned hybrid DNA fragments of the individual under investigation.

As described in the RFLP section (see above) recombinant DNA is first made and then incorporated in a culture of *E. coli* which is plated on a petri dish of agar. The bacterial cells begin to reproduce, forming a lawn of cells on the surface of the agar. Any *E. coli* cells that are successfully 'infected' by the recombinant DNA molecules will manufacture many new copies of the recombinant DNA, and they appear on an agar plate as a clear circular plaque in the lawn of cells, see Fig. 12.4. Each plaque represents a clone of the original recombinant DNA molecules and can be propagated indefinitely by infecting more *E. coli* cells. Although each plaque produces a unique clone of recombinant DNA molecules, *which* piece of DNA is present in a given plaque is entirely a matter of chance, since all the genetic material of donor and recipient will be present, even if only one particular restriction enzyme has been used for the cutting.

The problem now is to find a plaque that contains the fragment of DNA in which one is interested. For example, has the original DNA come from a person carrying the thalassaemia gene or from an unaffected, (or non-carrier) relative? For this we need the probe, already described. It is used as follows, and the procedure is known as the Southern blotting technique (after Dr Southern, originally of Edinburgh). A sheet of filter paper made of nitrocellulose, to which DNA binds strongly, is gently pressed against the surface of each agar plate containing the plaques. Some of the DNA in each plaque is absorbed by the filter. The absorbed DNA is then denatured i.e. dissociated into two single strands and the filter is bathed in a solution containing the radioactive complementary DNA molecules – the probes. If this radioactive probe encounters complementary strands of DNA it binds to them by base pairings – it has 'sought out and found'. The filter is then pressed against a sheet of photographic film. Any exposed areas that develop on the film represent plaques containing the cDNA molecules combined with the cDNA on the probe. Because the original plaques derived from the individual being investigated have been undamaged by this procedure, more molecules of this particular sample of cDNA can be propagated in additional *E. coli* cells to produce as many copies of the sample as we wish. Figures 12.4 and 12.5 show diagrammatically the events that have been described and it will be seen that the patient has in fact the disease of thalassaemia.

It might be asked why it is necessary to have such vast quantities of cloned fragments. There are several reasons for this:

(*a*) if one does not have a great many copies one is looking for a needle in a haystack – the gene being sought represents only one millionth of the DNA in a cell;

(*b*) many copies are needed for analyzing base sequences and the inserted

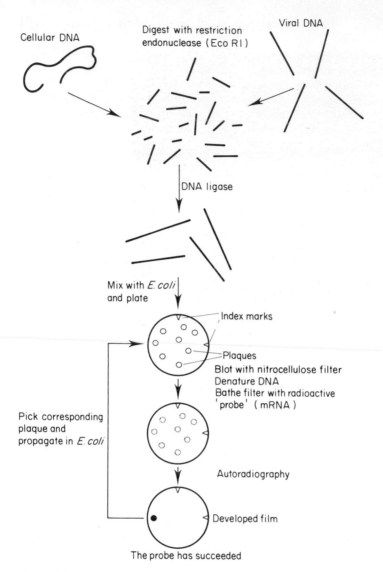

Cellular DNA

Digest with restriction
endonuclease (Eco RI)

Viral DNA

DNA ligase

Mix with *E. coli*
and plate

Index marks

Plaques
Blot with nitrocellulose filter
Denature DNA
Bathe filter with radioactive
'probe' (mRNA)

Pick corresponding
plaque and
propagate in *E. coli*

Autoradiography

Developed film

The probe has succeeded

Fig. 12.4 How a gene probe is used. Viral DNA is cleaved by a restriction endonuclease that produces staggered cuts so that single-stranded sticky ends are formed at each end. The same enzyme is used to prepare a sample of foreign DNA e.g. from an at risk individual. The pieces of DNA are then joined with the aid of the complementary bases of their sticky ends and DNA ligase. The recombinant DNA molecules are mixed with living *E. coli*, the host of the virus. Those recombinant molecules that have regained their infectivity kill their host cells, forming plaques in the 'lawn' of uninfected cells. Blotting the surface of the plate transfers some of the DNA from each plaque to the filter. After denaturing the blotted DNA to make it single-stranded, the filter is exposed to radioactive RNA (the probe) carrying the gene sequence you are looking for in the patient. Autoradiography reveals which, if any, plaques contain that DNA (Fig. 12.5). The original positive plaque(s) can be identified and more of its DNA used to infect *E. coli*. In this way, unlimited amounts of a particular gene can be produced. (Adapted from Kimball, 1983.)

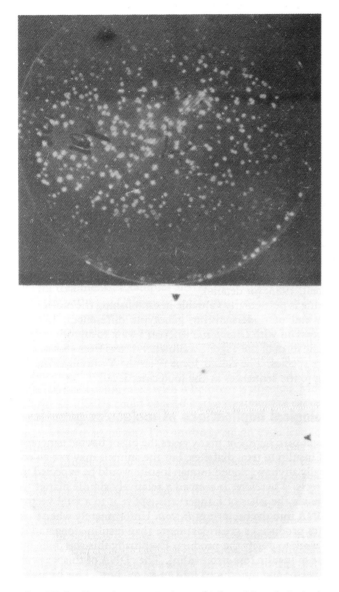

Fig. 12.5   *Top*: plaques on a lawn of infected *E. coli*. Each plaque represents a site of infection by viral DNA that may include DNA sequences from an at risk individual. Radioactive RNA (e.g. mRNA) is used as a 'probe' to detect plaques containing DNA sequences to which the probe can bind by base pairing. *Bottom*: one plaque has bound the radioactive probe strongly (upper right quadrant). Note the index marks on the autoradiogram which make it possible to indicate which of the plaques is the positive one. (Courtesy of Dr Ronald W. Davis.)

producing the required antibody it can be cultured indefinitely as an established cell culture.

A Rhesus anti-D (see p. 81) monoclonal antibody would be extremely helpful but there is some way to go before this is achieved. Not the least of the problems is to make sure that the prophylactic injection (which is given to healthy mothers) is entirely free from contamination by virus.

12.6   The use of a monoclonal antibody relevant to molecular genetics

A labelled monoclonal antibody can be used in conjunction with DNA technology to identify some problems in the brain, e.g. to discover the role of various peptides specific for brain tissue whose function is at present unknown but which might be relevant to diseases such as schizophrenia. One way to advance this theory is to extract the total mRNA from brain tissue and then produce a brain cDNA 'library'. With mRNA probes from different tissues, including brain tissue, it becomes possible to identify those clones which are brain-specific. These can be sequenced and the related peptides then synthesized. Using monoclonal antibodies to these peptides, tagged with a suitable marker, it would be possible to study their distribution within the brain – and this might be of great value in medical research – there is a hint, for example, that the lesions in schizophrenia are on the left side of the brain.

12.7   Conclusions

Clearly molecular genetics is with us to stay but it is becoming increasingly difficult to comprehend and the only solution seems to be for those in the VIth form, together with the masters and writers of books, to take a course in the discipline so that its bench work can be understood, particularly in order to differentiate between what has to be inferred and what can actually be seen at varying magnifications.

Transfection (see glossary) gathers way as we write. The gene responsible for luciferase, the firefly's lantern, can be incorporated into plant cells so that their internal structure can be better shown up on photographic films. (Times Science report, 4th August 1986.)

Glossary

Adrenergic blocker   A drug which acts via the autonomic nervous system, blocking the production of adrenalin or noradrenalin.

Allelomorph (allele)   One of two or more contrasted genetic characters found at the same locus on homologous chromosomes.

Antibody   A serum protein produced in an animal when a certain kind of substance (an antigen q.v.) which is normally foreign to its tissues gains access to them. The antibody reacts specifically with the antigen.

Antigen   Any substance capable of stimulating the formation of an antibody (q.v.) with which it reacts specifically.

Autosome   Any chromosome other than the sex chromosomes (X and Y). There are 22 pairs of autosomes in man.

Bacteriophage   A virus which infects bacteria.

B cells   Lymphocytes which secrete antibodies. (B stands for bone-marrow derived.)

Barr body   Small darkly staining body under the nuclear membrane of mammalian somatic cells present in normal females but absent in normal males (see p. 58).

Cistron   That portion of the chromosome in which the loci are integrated for one function, i.e. the smallest unit of genetic material which is responsible for the synthesis of a specific polypeptide.

Concordance (and discordance)   Both twins exhibiting the same trait (*discordance*: the opposite).

Coombs' test   A test which detects the presence of an antibody on the surface of the red cell.

Crossing-over (syn. recombination)   The exchange of genes between homologous chromosomes which takes place at meiosis.

Deletion   Loss of part of a chromosome, or loss of a DNA sequence.

Denaturation   A process by which the two strands of the double DNA helix are disrupted and dissociated into two single strands.

Dizygotic twins   These result from the fertilization of two separate eggs at the same time and two such twins are no more genetically similar than sibs. Another type of twin could be produced if two halves of an ovum were fertilized by two different spermatozoa. If this happened the twins would have an identical set of genes from the mother but a different set from the father.

Dominance   A character is said to be dominant if the gene controlling it produces the same effect in the heterozygous as in the homozygous state. Loosely, however, for clinical purposes, a condition is said to be dominant if it appears in heterozygotes. In many instances the homozygote has never been observed.

Drift   (see *genetic drift*)

Expressivity   The degree to which the effect of a gene is expressed. If the gene is controlling a disease some of those inheriting it will be more severely affected than others e.g. in neurofibromatosis some individuals will have skin tumours, pigmentation and bone changes, whereas others will have pigmentation only.

First degree relative   Parents, children, brothers and sisters, i.e. one's closest relatives.

Fitness (syn. biological fitness)   The fitness of an individual is measured by the number of his or her offspring who reach reproductive age. An individual is said to have unit fitness if he or she has two such offspring (not one, as each child must have two parents).

Gene library   (see *library*)

Genetic drift   The establishment of certain gene frequencies in small populations – not owing to natural selection but owing to the original genetic constitution of the ancestors of the population, or chance survival when a population is reduced.

Genome   The complete make-up of an individual–i.e. the entire set of nuclear genes, whether in a haploid or a diploid cell.

Genotype   The genetic make-up of an individual with regard to a given pair of alleles – a blood group A individual may be of a genotype AA or AO (cf. phenotype, q.v.).

Haplotype   The particular five HLA genes carried by an individual on each of his two number 6 chromosomes.

Hardy-Weinberg equilibrium   Unless disturbed by outside influences, e.g. natural selection, the proportion of the various genotypes in the population remains the same in each successive generation provided that the particular genes carried by an individual do not influence the choice of mate – i.e. mating is random (see p. 31).

Heterosis   Increased vigour of growth, fertility, etc., in a cross between two genetically different lines, due to greater heterozygosity.

Heterozygous   Possessing two different allelomorphs at the two corresponding loci on a pair of homologous chromosomes.

Homologous chromosomes   Chromosomes which are homologous pair with each other at meiosis and contain identical sets of loci (q.v.).

Homozygous   Possessing similar allelomorphs at the two corresponding loci on a pair of homologous chromosomes.

Incidence (cf. prevalence)   Number of new cases occurring during a specified period of time. The incidence rate is this number per specified unit of population, e.g. the number of new cases which occurred per year per thousand of the population.

Inversion (chromosomal)   Because of aberrant cross-overs within chromosomes a segment can become inverted and the genes will then appear in the wrong order.

Karyotype   The artificial arrangement of the chromosome set so as to allow comparison of the morphology, and subsequent analysis.

Lesion   Any morbid change in the function or texture of an organ.

Library (gene library)   Set of cloned DNA fragments which together represent the entire genome of a particular tissue.

Linkage   Genes situated on the same chromosome are said to be linked. Except when crossing-over (q.v.) occurs they are inherited together. Crossing-over occurs less frequently the nearer together the genes are situated.

Linkage disequilibrium   The association of two linked alleles more frequently than would be expected by chance.

Locus   The site on a chromosome occupied by a particular gene, or by a member of a particular allelomorphic series.

Lyonization   According to Lyonization, in mammals one of the two X chromosomes in every female somatic cell is inactive, this sometimes being the maternal and sometimes the paternal one. Activity or inactivity is decided shortly after the zygote is formed. Thereafter all the descendants of that cell resemble it.

Metastasis   A secondary growth of a malignant tumour at a site distant from the primary growth, from which cells have been carried by blood or lymph.

Monosomy Here one chromosome of an homologous pair is missing, hence the individual only has 45 chromosomes.

Monozygotic twins These result from the division into two of the embryo derived from a single fertilized ovum and such twins are genetically identical.

Mosaic An individual with cell lines, or tissues, of different chromosome numbers or constitutions.

Multifactorial inheritance (see p. 15)

Mutation Sudden change, either in an individual gene (point mutation) or in the structure of a chromosome (e.g. a translocation).

Non-disjunction The failure of two homologous chromosomes to pass into separate gametes either at meiosis or mitosis. Both therefore pass to the same daughter cell.

Penetrance A gene controlling a dominant character is said to have full penetrance when the character it controls is always evident in an individual possessing the gene. A gene controlling a recessive character is said to be fully penetrant if the character is invariably manifest when the genes are present in double dose.

Penetrance = frequency with which any effect is shown in a population.

Expressivity = degree to which effects are shown in an individual.

Phenocopy An exact replica of an inherited character but produced environmentally.

Phenotype The manifest genetic make-up of an individual, e.g. the information available from the examination of a single individual, without reference to any family (breeding) data. (cf. genotype q.v.)

Plasmid A genetic element consisting of circular DNA, capable of replication within a bacterium.

Pleiotropy Multiple effects of the same gene. A gene might have a major effect such as that controlling the production of a blood group antigen and also play a minor part in predisposing to duodenal ulcer, for instance.

Polymorphism The occurrence within a freely interbreeding species of widely differing inherited forms, the rarest of them being too common to be kept in existence by recurrent mutation. The term has a slightly different meaning in genetic engineering terminology.

Prevalence (cf. incidence) Total number of cases of a disease existing in the population at a specified time. The prevalence rate is this number expressed per thousand individuals.

Primed Immunized, but antibodies not detectable except by special methods such as cell survival studies.

Prion A proteinaceous infectious particle containing neither DNA nor RNA.

Probe (see Chapter 12, p. 102)

Propositus The individual through whom the investigation of a pedigree is begun; usually, but not always, an 'affected' individual. (*syn. proband*)

Radioimmunoassay A method for the microassay of antigenic molecules. It is based on competition between radioactively labelled standard solutions of antigen for a fixed quantity of antibody. The antigen–antibody complex is then assayed.

Restriction enzyme (see Chapter 12, p. 100)

Reticulocyte Non-nucleated precursor of a red blood cell.

Rhesus blood group system In 1940 Landsteiner and Wiener, having immunized rabbits and guinea-pigs with the blood of the monkey *Macacus rhesus*, made the very surprising discovery that the resulting antibodies agglutinated not only the monkey red cells but also the red cells of about 85% of white people in New York. The 85% whose red cells were agglutinated by the rabbit anti-rhesus serum the authors called Rh-positive, the remaining 15% Rh-negative.

Recessive A character is said to be recessive if it is only manifested in those homozygous for the gene controlling it. It is not detectable (except sometimes by special tests) in the heterozygote.

Recombination   (see *crossing-over* – but used in another sense in genetic engineering)

Robertsonian translocation   (see *translocation*)

Sensibilized   Immunized, but with antibodies only detectable by special tests.

Sex-linkage   A gene is said to be sex-linked when it is on part of either the X or the Y chromosome. If it is situated on the non-pairing part of the Y it can never cross-over on to the X and will therefore always be handed from father to son. If a gene is on the X a man will pass it on to all his daughters and a woman to either son or daughter.

Supergene   Term used to denote a series of genes which have become closely linked on the same chromosome owing to the selective advantage of their being inherited as a unit; crossing-over can, however, occasionally occur.

T cells   Lymphocytes concerned with cellular immunity, i.e. the reaction to a nettle sting, or to an organ transplant. (T stands for thymus-derived)

Transfection   The introduction by micro-injection of foreign DNA into the nuclei of early embryos of mice. By doing this it has been found possible to restore growth in a dwarf mutant mouse, the result of an inherited deficiency in growth hormone. Human growth hormone fused to a heavy metal was introduced into mutant mouse eggs. The transgenic dwarf mice grew to approximately 1.5 times the size of heterozygous animals. Most important of all, the fusion gene was integrated into the host chromosome and transmitted to the next two generations. (Hammer *et al.*, 1984.)

Translocation   This happens when two pieces of a chromosome are broken off from two non-homologous chromosomes and their positions exchanged (this is reciprocal translocation). If one of the broken pieces is small this is usually lost and the result is then known as a Robertsonian translocation.

Triplet   A series of three bases in the DNA or RNA molecule which codes for a specific amino acid.

Trisomy   Where one chromosome is represented by three homologues as opposed to the normal two. Hence the individual has 47 chromosomes.

Xg blood group   A blood group system which is sex-linked, i.e. the genes controlling the characters are on the X chromosome. In the original series tested 61.7% of men were Xg(a+), while 38.3% were Xg(a–). For women the percentages were Xg(a+) 88.8 and Xg(a–) 11.2 The reason for the difference between the male and female percentages is that some of the Xg(a+) women are heterozygous (Race and Sanger, 1975, *Blood Groups in Man*, 6th Ed., Oxford).

Zygote   The fertilized egg (ovum).

References

Alberts, B., Bray, D., Lewis, J., Raff, M., Roberts, K., Watson, J. D., (eds) (1983). *Molecular Biology of the Cell.* Garland, New York.

Allison, A. C. (1954). Protection afforded by the sickle-cell trait against subtertian malarial infection. *British Medical Journal*, **i**, 290–4.

Bailey, N. T. J. (1959). *Statistical Methods in Biology.* English Universities Press, London.

Barker, Jnr, R. H. *et al.* (1986). Specific DNA probe for the diagnosis of *Plasmodium falciparum* malaria. *Science*, **231**, 1434–6.

Buckwalter, J. A. and Tweed, G. V. (1962). The Rhesus and MN blood groups and disease. *Journal of the American Medical Association*, **179**, 479–85.

Carter, C. O. (1962). *Human Heredity.* Penguin, Harmondsworth.

Clarke, C. A. (1964). *Genetics for the Clinician*, 2nd edition, Blackwell Scientific Publications.

Clarke, C. A. (ed.)(1975). *Rhesus Haemolytic Disease.* Medical and Technical Publishing Company, Lancaster.

Clarke, C. A., Donohoe, W. T. A., McConnell, R. B., Martindale, J. H. and Sheppard, P. M. (1962). Blood groups and disease: Previous transfusion as a potential source of error in blood typing. *British Medical Journal*, **i**, 1734–6.

Clarke, C. A., Edwards, J. Wyn, Haddock, D. R. W., Howel Evans, A. W. McConnell, R. B. and Sheppard, P. M. (1956). ABO blood groups and secretor character in duodenal ulcer. Population and sibship studies. *British Medical Journal*, **ii**, 725–31.

Cornell, J. *et al.* (1983). Neural tube defects in the Cape Town area, 1975–1980. *South African Medical Journal*, **64**, 83–4.

Curtis, J. J. *et al.* (1983). Remission of essential hypertension after renal transplantation. *New England Journal of Medicine*, **309**, 1009.

Danks, D. M., Allan, J. and Anderson, C. M. (1965). A genetic study of fibrocystic disease of the pancreas. *Annals of Human Genetics*, **28**, 323–56.

Emery, A. E. H. (1983). *Elements of Medical Genetics*, 6th edition. Churchill Livingstone, Edinburgh.

Emery, A. E. H. (1984). *An Introduction to Recombinant DNA.* John Wiley, Chichester.

Evans, D. A. P., Manley, H. K. and McKusick, V. A. (1960). Genetic control of isoniazid metabolism in Man. *British Medical Journal*, **ii**, 485–91.

Fisher, R. A. (1930). *The Genetical Theory of Natural Selection.* Oxford.

Ford, E. B. (1940). *The New Systematics.* Huxley, J. (ed.). Oxford.

Ford, E. B. (1965). *Genetic Polymorphism.* All Souls Studies. Faber and Faber, London.

Ford, E. B. (1973). *Genetics for Medical Students*, 7th edition. Chapman and Hall, London.

Hamilton, M. Pickering, G. W., Fraser Roberts, J. A. and Sowry, G. S. C. (1954). The aetiology of essential hypertension. *Clinical Science*, **13**, 11–37.

Hammer, R. E., Palmiter, R. D. and Brinster, R. L. (1984). The introduction of metallothionein-growth hormone fusion genes into mice: In *Advances in Gene Technology: Human Genetic Disorders*, Vol I, 52–5, Scott *et al.* (eds). ICSU Short Reports, Cambridge University Press.

Harper, P. S. (1981). *Practical Genetic Counselling.* Wright, Bristol.

Harris, H. (1970). *Cell fusion.* University Press, Oxford.

Harris, H. (1971). Cell fusion and the analysis of malignancy. *Proceedings of the Royal Society of London*, B, **179**, 1–20.

Herndon, C. N. (1962). Empiric risks. In *Methodology in Human Genetics*. Burdette, W. J. (ed.). Holden-Day, San Francisco.

Khajii, T. and Ohama, K. (1977). Androgenetic origin of hydatidiform mole. *Nature*, **268**, 233–4.

Kimball, J. W. (1983). *Biology*, 5th edn. Addison Wesley Publishing Co., Massachusetts.

Lawler, S. D. and Sandler, M. (1954). Data on linkage in Man: elliptocytosis and blood groups, IV, families 5, 6 and 7. *Annals of Eugenics (Camb).*, **18**, 328–34.

Lyon, M. F. (1961). Gene action in the X-chromosome of the mouse (*Mus musculus* L.). *Nature.*, **190**, 372–3.

Maynard Smith, J. (1982). *Evolution Now.* London.

McKusick, V. A. (1985). The Human Gene Map 1 December 1984: the morbid anatomy of the human genome. *Clinical Genetics*, **27**, no. 2.

Miall, W. E. and Oldham, P. D. (1958). Factors influencing arterial blood pressure in the general population. *Clinical Science*, **17**, 409–44.

Pembrey, M. E., Winter, R. M. and Davies, K. E. (1985). A premutation that generates a defect at crossing-over explains the inheritance of fragile X mental retardation. *American Journal of Medical Genetics*, **21**, 709–17.

Penrose, L. S. (1959). Natural Selection in Man; some basic problems. In *Natural Selection in Human Populations*. Pergamon Press, London.

Pickering, G. W. (1959). The nature of essential hypertension. *Lancet*, **ii**, 1027–8.

Platt, R. (1959). The nature of essential hypertension. *Lancet*, **i**, 55–7.

Platt, R. (1963). Heredity in hypertension. *Lancet*, **i**, 899–904.

Race, R. R. and Sanger, R. (1975). *Blood Groups in Man*, 6th edn. Blackwell Scientific Publications, Oxford.

Renwick, J. H. and Lawler, S. D. (1955). Genetical linkage between the ABO blood group and nail-patella loci. *Annals of Eugenics (Camb.)*, **19**, 312–19.

Renwick, J. H. and Schultze, J. (1965). *Annals of Human Genetics*, **28**, 379–92.

Roberts, J. A. Fraser (1973). *An Introduction to Medical Genetics*, 6th edn. Oxford University Press, London.

Roberts, J. A. Fraser, and Pembrey, M. E. (1978). *An Introduction to Medical Genetics*, 7th edn. Oxford University Press, London.

Sheppard, P. M. (1975). *Natural Selection and Heredity*, 4th edn. Hutchinson, London.

Smithells, R. W. *et al.* (1981). Apparent prevention of neural tube defects by periconceptional vitamin supplementation. *Archives of Disease in Childhood*, **56**, 911–18.

Søbye, P. (1948). *Heredity in Essential Hypertension and Nephrosclerosis*. Ejnar Munksgaard, Copenhagen.

Swales, J. D. (ed.) (1985). *Platt versus Pickering*. The Keynes Press.

Tovey, L. A. D. (1984). The contribution of ante-natal anti-D prophylaxis to the reduction of the morbidity and mortality in Rh hemolytic disease of the newborn. *Plasma Therapy Transfusion Technology*, **5**, 99–104.

Turner, G. *et al.* (1978). Marker X chromosomes, mental retardation and macroorchidism. *New England Journal of Medicine*, **219**, 1472.

Warr, J. R. (1984). *Genetic Engineering in Higher Organisms*. Studies in Biology no. 162. Edward Arnold, London.

Weatherall, D. J. (1986). *The New Genetics and Clinical Practice* 2nd edition. Oxford Medical Publications.

Index